THE INVENTOR'S COMPLETE HANDBOOK

How to Develop, Patent, and Commercialize Your Ideas

By James L. Cairns, Ph.D.

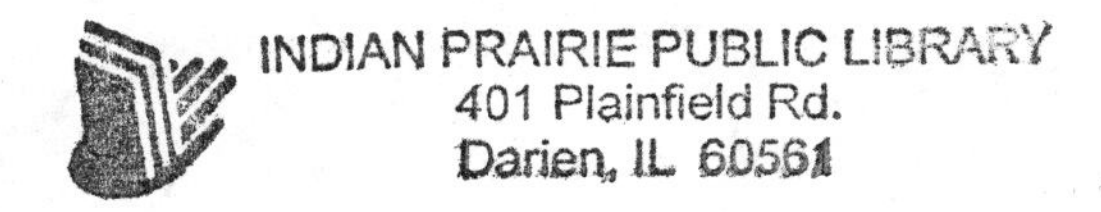

THE INVENTOR'S COMPLETE HANDBOOK: HOW TO DEVELOP, PATENT, AND COMMERCIALIZE YOUR IDEAS

Previously Published in 2006 as The Inventor's Pathfinder — Now Updated and Expanded

Atlantic Publishing Group, Inc.
1405 SW 6th Avenue • Ocala, Florida 34471 • Phone 800-814-1132 • Fax 352-622-1875
Web site: www.atlantic-pub.com • E-mail: sales@atlantic-pub.com
SAN Number: 268-1250

Library of Congress Cataloging-in-Publication Data

Cairns, James L., 1937-

The inventor's complete handbook : how to develop, patent, and commercialize your ideas / James L. Cairns.

pages cm

Includes bibliographical references and index.

ISBN 978-1-62023-018-3 (alk. paper) -- ISBN 1-62023-018-6 (alk. paper) 1. Inventors--Handbooks, manuals, etc. 2. Patents--Handbooks, manuals, etc. I. Title.

T212.C35 2015

600--dc23

2015004223

Printed in the United States

Printed on Recycled Paper

Reduce. Reuse. RECYCLE.

A decade ago, Atlantic Publishing signed the Green Press Initiative. These guidelines promote environmentally friendly practices, such as using recycled stock and vegetable-based inks, avoiding waste, choosing energy-efficient resources, and promoting a no-pulping policy. We now use 100-percent recycled stock on all our books. The results: in one year, switching to post-consumer recycled stock saved 24 mature trees, 5,000 gallons of water, the equivalent of the total energy used for one home in a year, and the equivalent of the greenhouse gases from one car driven for a year.

Over the years, we have adopted a number of dogs from rescues and shelters. First there was Bear and after he passed, Ginger and Scout. Now, we have Kira, another rescue. They have brought immense joy and love not just into our lives, but into the lives of all who met them.

We want you to know a portion of the profits of this book will be donated in Bear, Ginger and Scout's memory to local animal shelters, parks, conservation organizations, and other individuals and nonprofit organizations in need of assistance.

– Douglas & Sherri Brown,
President & Vice-President of Atlantic Publishing

Disclaimer

There are many Internet sites referenced in this book. They are current at the time the book is written; however, they change from time to time so some effort might be needed to track them down as time goes on.

This book provides general advice and guidance for inventors. Care has been taken to ensure that the information it contains is accurate and up to date; nevertheless, errors and omissions are possible, and some might still be lurking in here somewhere. To be certain you have the best, most accurate advice for critical actions such as legal matters, you should seek personalized, professional assistance from a lawyer licensed to practice in your state.

TABLE OF CONTENTS

T. A. EDISON.
Electric-Lamp.

No. 223,898. Patented Jan. 27, 1880.

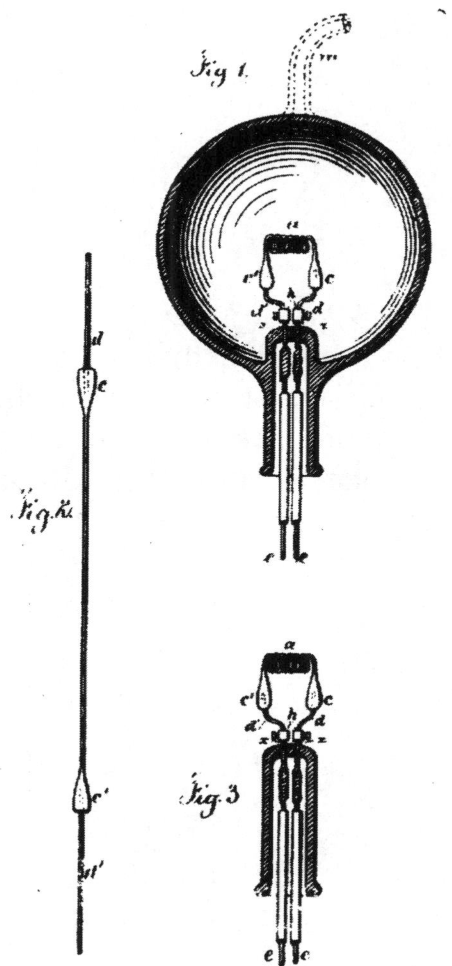

Inventor
Witnesses
Thomas A. Edison
Chas H Smith
Geo T Pinckney
per Lemuel W. Serrell
atty.

"I do not think there is any thrill that can go through the human heart like that felt by the inventor as he sees some creation of the brain unfolding to success... Such emotions make a man forget food, sleep, friends, love, everything."

— *Nikola Tesla*

ACKNOWLEDGMENTS

Over the years many people have helped and guided me in my work, and thus are important contributors to this book. Some of them are:

- Annamaria Negri Cairns, my life partner who has helped me pursue my inventing adventure for more than 25 years;
- Jackie Barrett, whose encouragement and support freed me in the early years to devote full time to inventing;
- Walter H. Munk, my mentor and longtime friend, who taught me to think clearly about abstract problems;
- John A. Folvig, Jr., who has been my partner in all of my major entrepreneurial adventures: and,

- Michael Vollmar, whose stamina and dedication were fundamental to our success.

I sincerely thank them all for their help. They have each been immensely important in both my life and my career.

I thank H. Michael Hartmann and Wesley O. Mueller for their valuable critical comments, and Connie Garzon-Bernal, Sonia Wadsworth, David Pfosi, Srikanth Ramasubramanian and Ned Harper for their help with the manuscript and its publication.

The Figure 1 Map was created by well-known San Diego artist Michael Dormer, now deceased.

PREFACE

Anyone with an entrepreneurial spirit who's willing to learn the essential techniques of inventing can practice the art successfully. The fundamentals of inventing, like those of any complex endeavor, must be acquired before they can be effectively applied, even by those of extraordinary natural talent.

This book should be the first read of an aspiring inventor. Many do-it-yourself books treat singular aspects of inventing such as patenting, and licensing of patent rights. They can be instructive, sometimes useful, in their particular areas. Studying this book first will put them in proper context relative to the whole inventing process. Once that process is understood, it will be clearer whether, when, and how to utilize do-it-yourself resources.

A significant part of what can be learned from this book does not reside between its covers. Instead, it is found in the many sources of supplemental in-depth educational material to which the reader is directed. A diligent student will acquire all the tools needed to be a competent independent inventor.

Creating and profiting from inventions is not the black art some imagine; it's one that can be mastered. Once you have acquired the techniques, you will have opened the door to a lifetime of exciting adventure. The path an independent inventor must follow to be successful is both arduous and fulfilling, as you will now learn.

Inventing begins with the recognition of a fundamentally good idea. This book teaches how to get that idea and what to do once you have it. Simply buying the book is not enough; unfortunately, you actually have to read it! It takes some hard work and careful study, but it's worth it.

Inventing is often mistakenly viewed as a three step process: get a good idea, patent it; then, cash in. In fact, those are essential steps, but there's much more to it than that. Novice inventors attempting to follow those overly simplistic steps are often stopped in their tracks by step two: the patent. Patents are expensive, and an unseasoned inventor stalls, uncertain if his innovation merits the investment. Or even worse, he goes ahead and patents a useless invention.

What's generally not realized is that there is a series of important steps to take between inspiration and investing in a patent. Most are necessary, and failure to follow them can have disastrous results. One critical step is to gain confidence in the invention. Another is to see if somebody else already owns it. Still another is to gauge its commercial potential. These are early, inexpensive parts of the process the inventor can do himself. And they're even fun! Following them, the inventor who is not quite ready to pay for filing a traditional patent application can still get one year's temporary protection by filing an inexpensive, provisional patent application. In that year the inventor can develop a pretty good feeling for an invention's value before investing significant money.

Once the invention is patented, there are additional careful actions to take in deciding how best to profit from it. All of the necessary major steps are described in this book.

You'll find that inventing is pretty much a solitary game. Others might contribute in important ways to the effort, but the inventor is the one who doesn't give up, who's there when everyone else has gone home, and who, against long odds, carefully finds a path through the labyrinth of impeding obstacles. Persistence is the grit that will get you through. When you're convinced you can do something, don't give up. Stick with it. Practically nothing works perfectly the first time. Countless immense successes, such as those of Edison, Marconi, Bell and many other famous inventors were steeped in early failure. There's far more to be learned from failure than from success, so don't be discouraged by it. It's an important part of inventing.

Believe in yourself. Believe in your ideas. Be frugal and work hard. This book will guide you and help you avoid costly mistakes. Good luck. You'll never be bored!

4 Sheets—Sheet 1.

L. J. McCORMICK, W R. BAKER & L. ERPELDING.

HARVESTERS.

No. 193,770. Patented July 31, 1877.

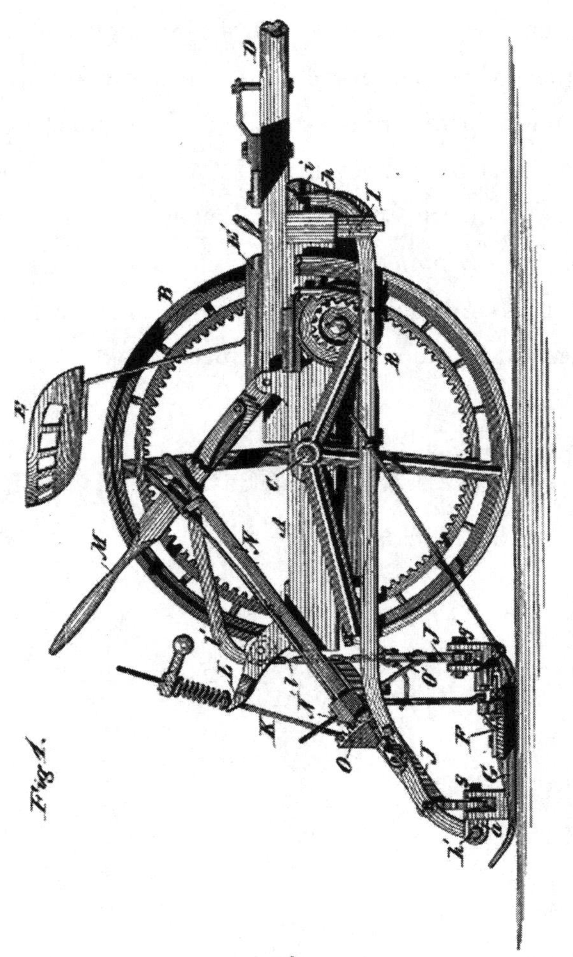

WITNESSES.

Harry King

Wm J. Peyton

INVENTOR

L. J. McCormick, Wm R. Baker & L. Erpelding

By

Wm D. Baldwin their Attorney

N. PETERS, PHOTO-LITHOGRAPHER, WASHINGTON, D. C.

PROLOGUE

"Everything that can be invented has already been invented"

— Charles H. Duell, director of the U.S. Patent Office, 1899

Charlotte spins her web with exquisite precision. She has carefully chosen a site with abundant prey, favorable natural elements, and sound anchor points. She tensions her web for optimum strength and sensitivity. She's making on-the-spot decisions, considering alternatives, and looking for ways to improve her success. She's inventing.

You, like Charlotte and every other creature, are already an inventor. You have to be, just to get through the challenges of a normal day. And sometimes, as you search for your own ways to overcome obstacles, you might find a solution with potential applications beyond your immediate needs, maybe even one with real commercial value. When you get that valuable, original idea you can profit from it, if you know how. But reaping financial rewards even from a great idea is difficult in the best of circumstances, and nearly impossible if not done properly.

This book presents successful techniques for maturing, protecting and commercializing innovations, and shows how to avoid the many pitfalls into which inexperienced inventors often sink. It pulls together a half-century of practical experience into a plain-language guide intended primarily for the beginner. The special skills needed for commercially successful inventing are not normally taught in schools, and there are no other books such as this one that provide a completely comprehensive treatment of the subject.

The inventing process is described herein as a journey from first concept to financial rewards. The route is along three intertwined pathways: technical, legal, and commercial. Along the technical portions of the route an invention matures from first thought to market-ready product. The inventor must, from time-to-time, travel through legal territory to define and protect his ownership, and to enter into contractual agreements. Along the commercial pathway the most marketable product configurations are determined, choices are made about selling out, taking on partners, seeking outside investment, and most of all, just the day-to-day decisions of business.

The technical, legal, and commercial pathways are all fraught with danger. The journey is long and sometimes painful; many who set out never make it to their final destination. Poor choices along the way can quickly lead to disappointment, humiliation, and financial ruin. Inexperienced inventors drop by the wayside by making simple, needless, blunders or by falling prey to unscrupulous predators. That all sounds pretty dismal, but it doesn't have to happen to you. As you will soon appreciate, once you learn the art, the trip is not all that menacing. And, it's a great adventure.

I had my first income as an inventor in the 1960s. In the intervening half-century I have travelled the inventing route many times. My motivation for writing this book is to guide you successfully through your own journey. As you go along, I will be your seasoned travelling companion, pointing out both troublesome areas and opportunities along the way.

Remember, it's the independent inventor working in humble circumstances who is at the heart of American industry. Think about our nation's greatest enterprises: Westinghouse, Ford, AT&T, General Electric, Microsoft, Apple, Hewlett-Packard, Lockheed-Martin, and a very, very long list of others. All were spawned modestly by one or two individuals with a passion for their ideas. Were it not for continued innovation from independent inventors, such major industries would not exist. The impact a single individual can have on the entire world is amazing. Never think the idea you are working on is small or insignificant. The inventors who started our great industries could not have imagined at the outset that they were changing the world, but they did.

How did they get started? The creations of our most prominent inventors, like those of tiny Charlotte, came from attempts to solve the problems that confronted them. We are all stimulated to improve the things around us, things we are intimately familiar with, things that bother us.

As you set out to develop your own innovations, you're going to be taking some very serious chances, and making important emotional and financial investments. You will be passionately pursuing a dream, drawn by satisfaction and unlimited potential rewards. Nothing could be more enticing.

Inventing is like any other complex pursuit; you must first learn how to do it. I urge you to study each section of this book carefully, even though at first glance some topics might not seem relevant to your needs. Be thoughtful in what you do, and you will have a safe journey.

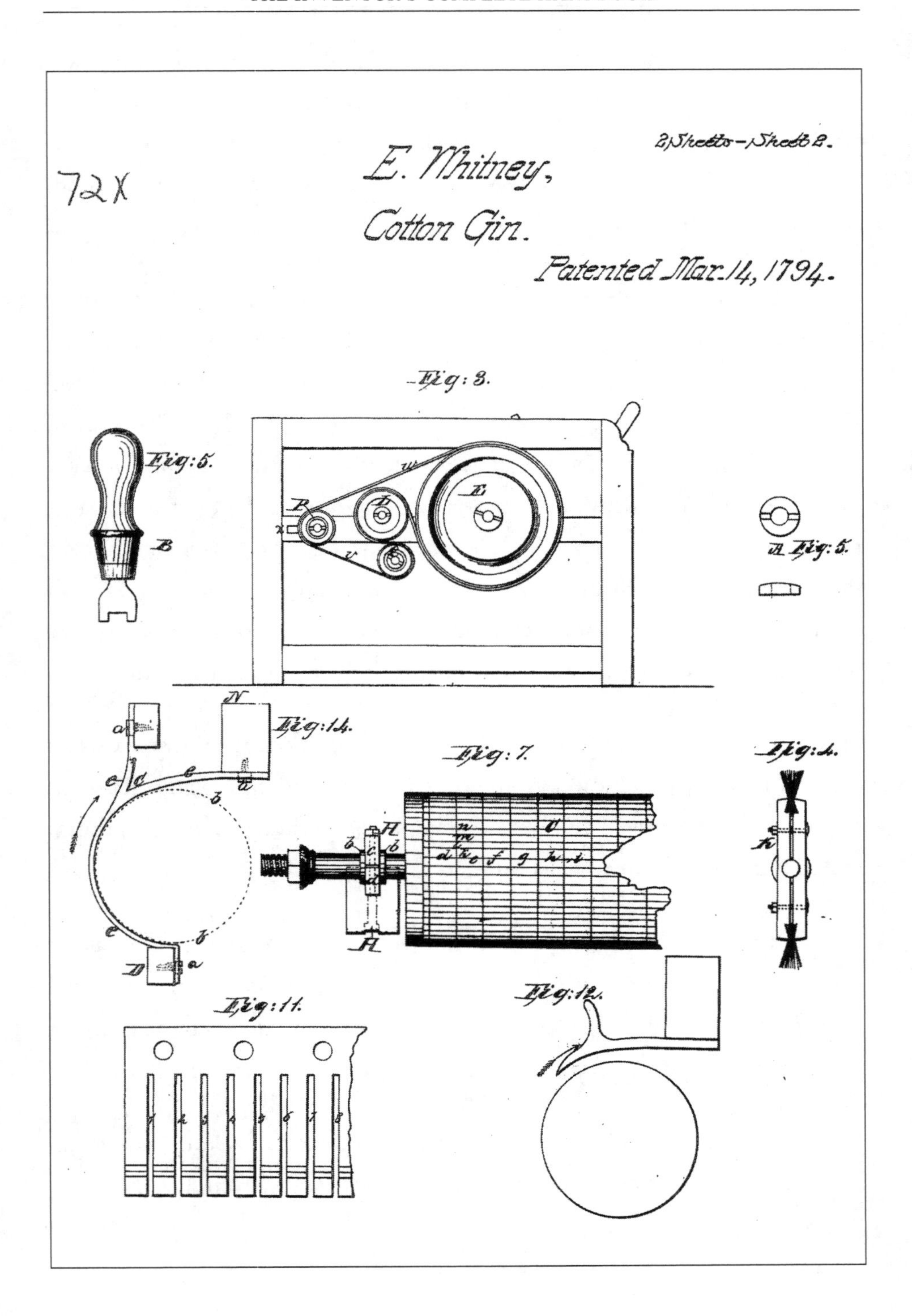
2 Sheets—Sheet 2.
72X
E. Whitney,
Cotton Gin.
Patented Mar. 14, 1794.
Fig: 3.
Fig: 5.
Fig: 5.
Fig: 14.
Fig: 7.
Fig: 4.
Fig: 11.
Fig: 12.

AN INTRODUCTORY OVERVIEW OF THE JOURNEY

There's no set way to pre-judge every turn on the path to success because every invention, by definition, is different from every other. Obviously some inventions are more complex than others. Even so, certain things simply must take place to transform any idea into a market-viable product. Regardless of who does them or what the product is, the essential steps are pretty much the same in every case.

Along the inventor's path, there are decision points where you'll have to choose how to go forward. You might consider cashing in by selling or leasing your ownership, for instance. Or you could stay the whole distance: developing, manufacturing, and ultimately selling the product; you might even sell the company you have created to manufacture it. Criteria for staying or getting out at various times will be examined as they arise. The choice will depend on your particular invention and your goals and circumstances at the time.

You can opt out anywhere along the way if you wish, but the invention itself must go through every step from start to finish if it's to get onto the market.

The path to be traveled is shown on the Map in Figure 1 and briefly described in the rest of this section. This is just a preview. We will study every element of the path in much greater detail as we go along in the book. As the Map implies, the trip is through rough terrain with danger lurking everywhere along the way. Do not be discouraged by that; the difficulties are mostly avoidable once you know how to navigate the route.

In the context of the Map you could compare learning successful inventing to learning how to drive in traffic. Suppose you learned how to control a car out in a field somewhere, but have never driven on streets. Now, you want to drive across a busy city. If you knew nothing about which lane to drive in, stop signs, pot holes, cross-walks, railroad crossings, speed limits, and many other essential factors, you would very likely get into serious trouble. Once you have that knowledge, though, your trip is relatively safe. Inventing is like that; once you know how, you can travel the route with greatly reduced risk. Now, if you study diligently, you will learn the rules.

You could view gates indicated on the Map as barriers through which your invention must pass to advance to the next step. If it cannot successfully pass through a gate, either the invention or the passing criteria might be modified to allow it through. If it simply cannot pass, the invention has failed and you're back to square one.

The Map is a very simplified version of the actual experience. Many critical decisions have to be made along each path segment. These will be discussed as we go, and the Map will serve to outline the overall direction. It also defines an order to the major product-development elements that should not be violated without serious consideration. Here's a quick overview of the route.

The Map

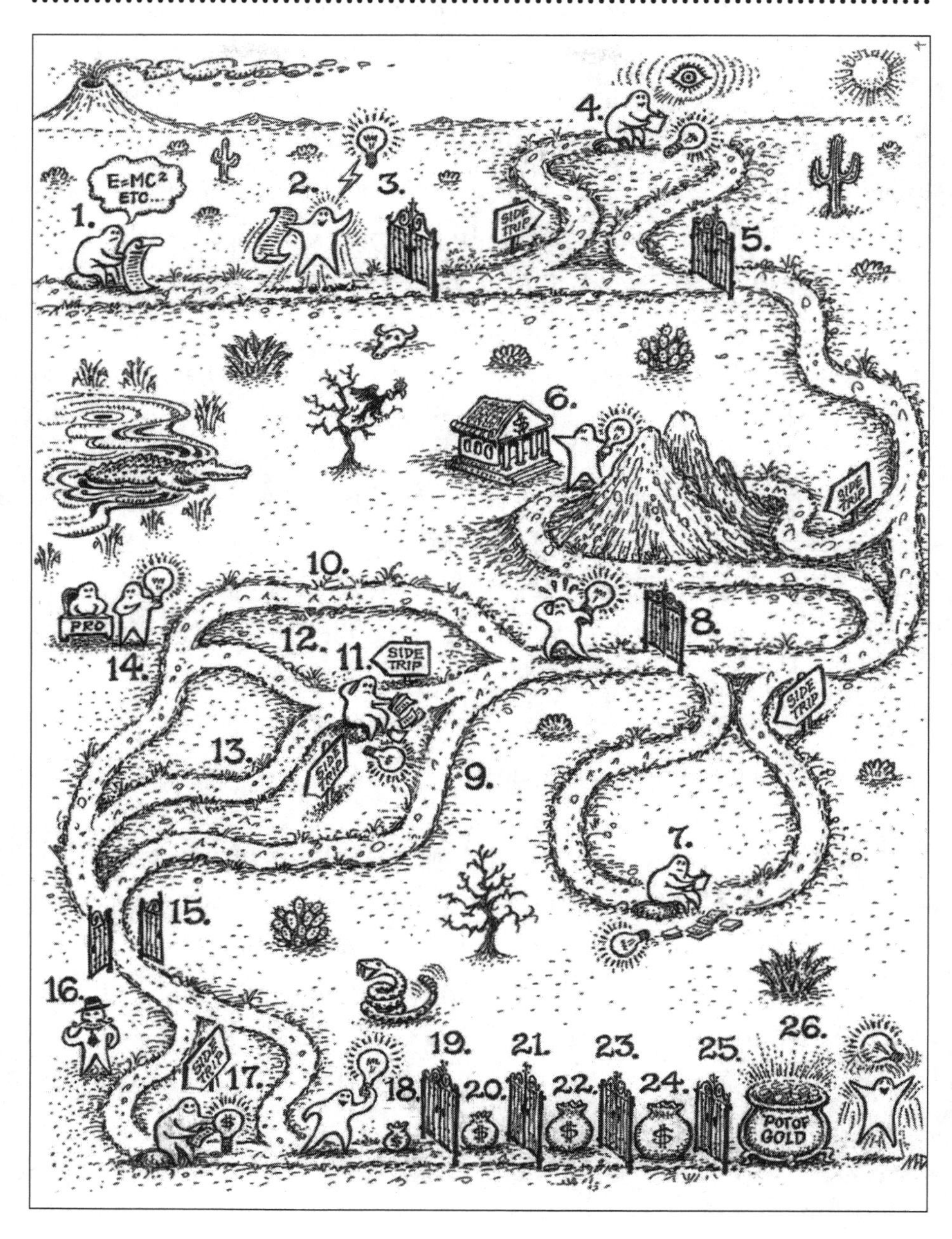

Figure 1

The inventing journey begins where you are right now, in a world filled with an infinite number of problems, needs, and desires. The first step, at point [1] on the Map[1], is to become sensitive to them, listing ones which interest you and which you might be able to resolve. After your interesting-problem list has grown a bit, you'll choose its most promising candidate to work on [2]. Do so carefully; you will be making a major commitment to your choice, shepherding it throughout the entire journey. Try to find a challenge for which you have a real passion. You will call upon that passion to keep you going when things get rough. Perhaps you already have picked an invention that excites you and seems worthy of your time and risk. In that case, you have completed steps [1] and [2]. That gets you through the gate at [3].

Having chosen the invention to pursue, it is vitally important that you define and refine it as much as possible within the confines of your imagination. Think long and hard about each aspect of it. The ability to visualize every detail of one's invention within the mind's eye is the inventor's most versatile tool. To keep it sharp it should be used often. There's an opportunity to take a side trip past [4] to learn more about mind's eye visualization. Take it, even if it seems like an unnecessary delay along your travels.

Once you have refined your idea as much as you can within your own imagination, you are ready to pass through the gate at [5] and proceed once again on your journey. Back on the main path you'll next weigh your invention against certain criteria to see if it seems both realizable and commercially promising. If it does, you will have identified a "fundamentally good idea"[2] that merits your commitment. If it doesn't; then it's back to step [1] for another try.

Once you have qualified a "fundamentally good idea" there's another side trip, this time to Map point [6]. It will allow you to get some free, helpful, professional

1 Numbers in brackets in the text refer to like numbers on the Map.

2 The term *"fundamentally good idea"* is used often throughout the book. It will be defined later.

business guidance. It might even yield some early-stage development funding. The possibility of funding depends upon the invention, and the funding might cost you something. Still, depending on your circumstances, it is well worth looking into. The side trip to [6] could very well be the one that gives your program the boost it needs to really get going.

I recommend that before disclosing your invention to prospective investors you file an inexpensive “provisional” patent application. You will learn much more about this sort of patent application further down the path. Whether or not you file a provisional patent application, at least get a signed non-disclosure agreement from each person to whom you reveal your ideas.

The journey’s next event is to evaluate your chances of owning the idea. Claiming and protecting an idea’s ownership is often a complicated matter, so once again a side trip passing Map point [7] is in order. On this detour you will study the basic ways to protect your creative innovations, such as how to patent your ideas. It will take you a while, but it’s really worth doing. If you want to be an inventor, you need to know the fundamental rules for protecting your creations; and you must be able to understand both your ownership boundaries and your competitor’s. In the end, you will almost certainly need to work with a professional in these matters; however, knowing your own way around the rules is very useful, and it gives you the flexibility to do a lot of the groundwork yourself. As we go forward, you will appreciate the advantages of that. Once you have determined that your chances of establishing ownership are good, you are ready to pass through the gate at Map point [8]. If the chances seem poor, back to step [1].

The things you will have studied up to Map point [8] in the book will help you decide which of the three alternative paths to follow going forward. Please look at the Figure 1 Map when considering the following alternatives: you could take pathway [9] and file your own patent application directly; or you could take path [10], stopping at [14] to have a professional do a patent search and file a patent application on your behalf. Another possibility is to take still another side trip past [11] to conduct a do-it-yourself search for patented inventions close to yours. After your DIY search you can either follow path segment [12] to professional help at [14] or you can take path [13] and file the patent application yourself. The relative merits of these paths will be discussed in detail later.

Once the patent application is filed, you are ready to proceed through the gate at [15,] which is now open for you.

Beware: as soon as your patent application is published, you will be assailed by offers from unscrupulous patent marketing and development firms [16]. They will be tracking you for most of the rest of your trip.

The invention's value continues to grow significantly as each subsequent gate is successfully passed. As you approach these final gates, there's one more side trip [17] to take. It leads you through the various common ways to cash in on your invention. Armed with that knowledge you can go forward to the final steps.

Logical points to consider reaping rewards appear on the Map as sacks of money along the route. Although sometimes there are earlier opportunities to do so, the best cash-in points occur after your patent application is filed. Once it's filed, you have the possibility of simply cashing in on the rights that the filed application affords you at [18]. Keep in mind here that we are still talking about a patent application, and not a granted patent. As we'll see later, your ownership is not fully defined until the patent is granted, which typically will be two or more years after filing the application. For the moment, though, let's assume that the patent will eventually be issued with substantial ownership rights. If you choose not to cash in at [18], you will next confront the gate at [19].

To pass through [19] you must complete the next major event along the way: proof-of-concept testing. It's just what the name implies; it proves that the various fundamental concepts of your invention work. Demonstrating the validity of your invention's basic premises increases its value and gives you the confidence to go forward. Completing this step opens [19] and presents the next logical cashing-in opportunity [20].

If you choose not to cash in at [20] you must next open the gate at [21]. To do so, you must complete the product-definition phase in which the markets and applications for which the invention will find use are determined. Once its projected uses are pinned down, the invention is adapted to them. That involves determination of manufacturing techniques, materials, size, robustness, price, product lifetime, and numerous other factors. They all dictate how the product will be made and sold. When the product is finally defined, not only has its feasibility been demonstrated, but it has also been tailored to a known customer base. Once again, its value has increased remarkably, and you have arrived at logical cashing-in point [22].

If you wish to go forward beyond [22] you must next unlock the gate at [23]. It opens once you have completed the last technical hurdle: qualification testing. Once the invention's various marketable embodiments are determined, prototypes and/or production units are rigorously tested to ensure that the product does what it's supposed to do.

The qualification tests prove that your product will meet or surpass all expected standards of utility, lifetime, ruggedness, and so on demanded in actual use. In addition to its previously demonstrated attributes, the product is now ready to be manufactured and is qualified for use in the targeted market. Its value has increased even further. Passing qualification testing presents the next logical point [24] to cash in, or you can go for the high-risk big bucks through the gate at [25] to produce and market your invention yourself.

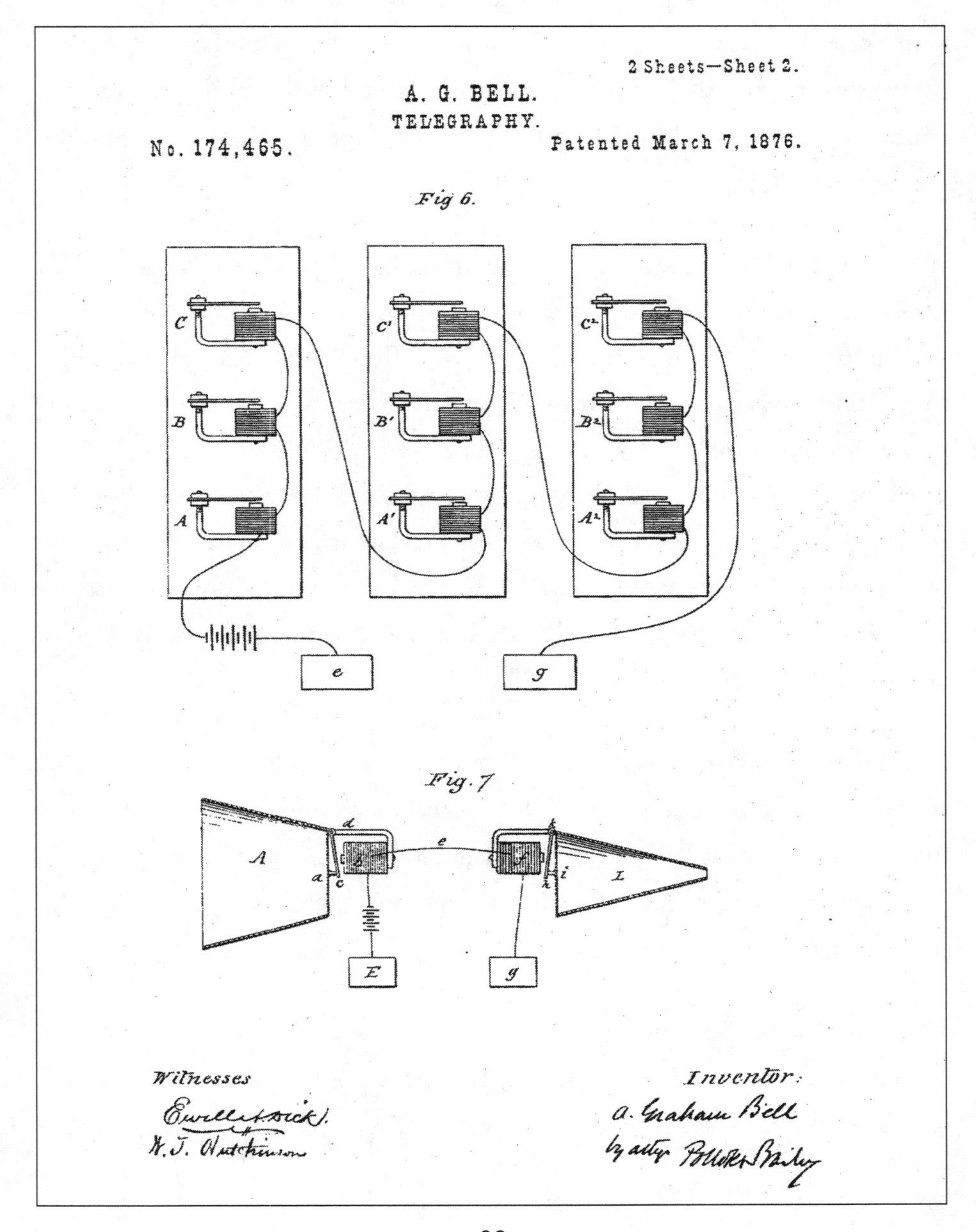

1 THE JOURNEY BEGINS

As with most endeavors, the first few steps weigh heavily toward success or failure. They are to be made very cautiously. The most critical step, and usually the first, is choosing what to invent; that is, what concept you want to bet your emotions, time and money developing. Getting "The Idea" is what most people think of as inventing, and, in fact, it's the keystone. But as we'll see, it isn't everything; it's just the beginning.

Here is the most important piece of advice in this book: Start out with a "fundamentally good idea." No one is a more biased judge of a new idea's value than the inexperienced inventor. The same unbridled enthusiasm that's essential to achieving success can blind a person to his[1] idea's shortcomings.

---

1 Where the gender of persons referred to in this book can be either masculine or feminine, the masculine is used by default. Please do not be offended by that. It is an expedient to avoid the awkward use of "he or she;" "him or her;" "himself or herself," and so on. The alternative of switching back and forth between masculine and feminine in the text is distracting and no less awkward.

Passionate optimism is an absolute requirement for an inventor, but unless alloyed with a trace of realism, it will sink him. The worst nightmare is to begin with an idea that doesn't have a prayer of success and then to doggedly stick with it. Persistence is fundamental to success; but before you start, be as sure as you can that you're persisting in a worthwhile endeavor. Staying with a tantalizing, really bad idea is a fatal, irreversible blunder. It is also one that is very commonly made.

So, how do you get that winning idea? It's true that occasionally inventions happen purely by accident. The most commonly cited example is Ivory Soap. Towards the end of the nineteenth century, Proctor & Gamble inadvertently mixed excess air into a production batch of soap, making it very lightweight. A short time later, customers began to request more of the "soap that floats." Discovery of a major product line that has endured for well over a century was entirely serendipitous. The manufacturers were astute enough to recognize their good luck, however, and took advantage of it.

History provides us with even more impressive accidents such as Fleming's discovery of penicillin, Roentgen's medical applications of X-rays[2], and Goodyear's use of vulcanized rubber. But alas, you cannot count on such good luck to bring you success. To get your great idea you're probably going to have to work for it.

Here's how to get started. The inventing process typically begins with the recognition of a solvable, unsolved problem. That doesn't often happen by luck, or by sitting around thinking: "Now I'll invent something; what in the world could it be?" Some stimulation is required. When you set out to invent something, don't start by trying to come up with a clever new product idea; that

2 Roentgen was awarded the Nobel Prize in Physics in 1901 for his development of x-ray diagnostic radiology. He was working with apparatus developed earlier for other purposes by Nicolai Tesla.

comes later. You must begin by looking for an attractive unsolved problem that interests you. Then see how you might creatively solve it. The first and most important challenge, then, is to find a really good problem, preferably one you passionately wish to solve.

The ability to recognize tractable problems is a powerful asset to an inventor. It allows him to choose a fertile starting point for his work. Those who repeatedly come up with good, marketable inventions have conditioned themselves to be on the alert for unsolved problems. Once they find a promising one, they look for its solution. Good problems are all around us. The trick is to recognize them. You'll be able to find them more easily with practice.

Here's some mental groundwork to help you become better at identifying them. Wherever you are as you read this book, look around you. Pick out some simple objects: a toothbrush, pencil, toy, or whatever. Try to visualize them within your mind's eye to figure out how they're made and how they work. How does the toothpaste get into the tube, for instance? Take simple things apart if that helps you. Imagine how they could be made better. Do this frequently and with a variety of different objects. It will increase your problem-solving capability. You'll also begin to more fully appreciate the thought that has gone into creating even the simplest objects in your world, and you'll become more astute at detecting shortcomings in those objects. The same sort of mental exercise can equally well be applied to non-mechanical inventions, such as electronic, pharmaceutical, software, chemical formulae, and so on. You'll get better at the exercises over time. Repeating them will improve your understanding of how things function and how they are made. That, in turn, will allow you to recognize ways to improve them.

With that same sort of critical thinking, as you go through your everyday routine, pay attention to the things that aren't quite right, that keep you from doing what you want to do, or that make it unpleasant. When that happens, stop and think how you might be able to improve whatever it is that irritates you. Try to completely understand the item or situation and think of what, if anything, about it can be done better. Usually if you've been bothered by things for a while, and you pay attention, you'll see ways to improve them. Through that sort of analytical thought will come the recognition of things that could be done easier, cheaper, better, or more reliably than they are now, or something useful that simply doesn't already exist. There you have a beginning concept for an invention. That's something you can start to work on. You'll be surprised by how many of them you'll come up with. Carry a small note pad and keep a list of bothersome things. Challenge yourself to add at least one new entry every day.

2 TIME TO CHOOSE YOUR BEST CONCEPT

Not all of the candidate concepts you find will be suitable for your first invention, so keep a record of your ideas and eventually choose the best one to focus on; very large, complex things should be avoided by a rookie inventor. That's not to say that if you have a great idea you should scrap it because it's complicated, but if you have such an idea it would be best to pursue it in cooperation with an entity that has the means to make it happen. The point is, if you plan to develop something yourself you should be pretty sure it's within your mental, physical, and financial capabilities. And above all, pick something that really excites you.

It's best to start out with things you can prototype at home from common hobby-shop or hardware store materials. That requires the maximum hands-on involvement on your part and the minimum expense, both of which are desirable. No matter what you choose to develop, everything will take longer, cost more, and be much, much more difficult than you first imagined. So, when you're looking around for a first thing to invent, think simple, clean, and small. After you've had a few successes you'll be able to afford grander mistakes.

Having chosen an invention to pursue, you are ready to pass through gate [3] on the Figure 1 Map. It's now time to focus upon advancing its development as far as you possibly can before committing any important resources other than your time. Of course you have been thinking hard about it since you first conceived it, but now you must really bear down to be sure you have done absolutely all that you can do to define it.

Begin by setting down your design goals; that is, list all the objectives you want your invention to accomplish. Consider them carefully. They're what will distinguish your product from those of your competitors. If your invention were a new sort of manual can opener, for instance, you might have the following four design goals, listed in order of importance:

1. Easily operated by both left- and right-handed people
2. Dishwasher safe
3. Compact
4. Non-corrodible

Once you've established your objectives, find as many ways to achieve them as you can. Rarely settle for the first way until you've found others with which to compare it. This cannot be stressed enough—don't stop when you've found a solution to your problem. Keep going; you'll find others. There's almost always more than one way to accomplish something, but only one best way. Collect all the possible solutions you can think of to solve your problem. In the end, sort through them, keeping the best.

Sometimes you'll find that one of your least important goals drives up the product's cost. In that case, you need to consider whether or not to lower your design expectations. Objective 4 in the above example, for instance, might require some components of high-priced, non-corrodible metals.

Perhaps that feature isn't important enough to your prospective customers to justify its inordinate cost. Therefore, as a matter of practicality you should lower your criterion for that design objective. It's very important for commercial reasons to choose your design goals in terms of not only what's possible or achievable but also what will sell at an acceptable price. A designer is often tempted to aim for the very best product that technology can provide, without regard for cost. It's wonderful to realize you've built a product of unsurpassable quality. But that euphoria will quickly turn to depression if no one's willing to pay for it.

Even though designing for practical economy just seems to be common sense, casual disregard for cost is probably the most common error made by inexperienced engineers and inventors. It is perhaps second only to almost completely ignoring assembly difficulties. The novice will rush to the most elegant rendition of his creation, certain that all will admire his creativity. Don't do that. Solid success is much more satisfying than extraordinarily clever failure.

Once you've determined the objectives that your invention should satisfy, it's time to begin defining its ideal embodiment. As you set out to solidify the ideal embodiment of your invention, you will be tempted to closely examine the inner-workings of existing, competitive products. Don't do it too soon. Looking too early leads you down design paths that have already been traveled, stifling your own creativity. The popular term "thinking outside the box" is relevant here. Looking too soon at what's already been done traps you inside a mental space with boundaries formed by existing concepts.

It's also best not to make detailed, formal drawings of your invention too early in the game. That, too, is a constraint. Maybe it's just laziness, but when a design's on paper, it's more tempting to modify it than to pitch it out and start with a completely new one. Drawings constrain your fluid creativity at a time when you should be free to mentally jump from one concept to another in a flash. The time for detailed drawings is when you have the innovation firmly in mind and can go no further through mental visualization.

That's not to say you shouldn't make notes and sketches along the way. It's essential to keep written, bound, and dated records from the start. They help you remember what you've done and document your progress. Your records might also be useful eventually in helping resolve ownership disputes. Carefully keep a bound journal of your work, but limit your drawings to simple sketches within it until your image is fully formed. For the moment, just try to do all of the work in your mind's eye.

The following section of the book offers a discussion of mind's eye visualization. It is not strictly part of the mechanics of inventing, but instead is a glimpse into the role the mind's eye plays, and how to possibly improve your imaginative powers.

Mind's eye Visualization: A Side Trip

We have the amazing ability to see the things around us in great detail. Our imaginations allow us to also "see" into the future to a certain limited extent. Here's an example: Imagine you are driving down the street in a quiet neighborhood, and a ball bounces out of the bushes and across the street well ahead of you. Immediately you apply the brakes, not because you are worried about hitting the ball, which has already passed, but because you know a child might soon follow. You have envisioned a possible future event based on the visual input received in the moment.

We also have some ability to "see" into the past. You might think about your favorite aunt, who passed away many years ago. You can see her in your mind's eye: see how she dressed, how she wore her hair, how she walked, talked, and so on. Maybe you cannot see her very clearly; the gift of sight within the mind's eye is not the same for everybody. Some of us have extremely acute mind's eye vision; others are blind, or nearly so. [1,2,3]

Fortunately for most of us, our mind's eye provides us with richly detailed visual experiences. In our dreams, for instance, we live extraordinary, lively episodes in which we see, feel, smell, and hear in exquisite detail the scenes in which we participate. And we feel all of the emotions of fear, joy, sorrow, and so on that go with them. They are very real to us as we dream them.

And then there's daydreaming. In a boring lecture, for instance, we might spend more time in the imaginary world than in the classroom. We defocus on the world around us as our thoughts drift off to imaginary places where we are momentarily living a different, more interesting, experience than the one in which our bodies reside.

Sleep-dreaming is completely passive, and for the most part daydreaming is too. When these happen they are not directed. Although they may be conditioned by recent events, we are not able to tune in a specific dream; the mind is sort of out of gear, and thoughts are freely roaming about.

1 For a glimpse into other's mind's eye abilities go to: http://intjforum.com/showthread.php?t=21067

2 The following website tells of a case of mind's eye blindness: http://discovermagazine.com/2010/mar/23-the-brain-look-deep-into-minds-eye.

3 Some questionnaires have been devised to give an indication of mind's eye visual capability. Marks, D.F. (1973). "Visual imagery differences in the recall of pictures". *British Journal of Psychology*, 64, 17-24.

In *The Inventor's Pathfinder*[4], a precursor to this book, it was proposed that the capability of mind's eye visualization could be improved with practice. That was not correctly stated. It is akin to saying "The more we practice looking with our eyes, the better our ocular vision will become." That's not right. It's more correct to say: "With practice we can improve our ability to use the mind's eye vision we already have."

Let me explain: In contrast to sleep-dreaming and daydreaming, the experience of imagining one's defunct aunt is not passive; we make a conscious choice to try visualizing her as she was, and to some extent we can do it. We could go further, and put a moustache on her, and have her dance a tango with a rose in her teeth if we wished. That is the creative employment of your mind's eye to visualize things that do not exist, and have never existed. That's what can be improved upon with practice. It can be directed; it can be used.

Although he is not generally well known in America, Nicola Tesla was perhaps the most gifted inventor of modern times[5]. He attributed a major part of his success to the method he used in approaching designs. He stated:

> *"My method is different. I do not rush into actual work. When I get a new idea, I start at once building it up in my imagination, and make improvements and operate the device in my mind. When I have gone so far as to embody everything in my invention, every possible improvement I can think of, and when I see no fault anywhere, I put into concrete form the final product of my brain."*

That is a great way to approach designing; however, it must be augmented with other factual information. Not all creative thought is visual. One of the principal ways rational thought takes place is verbally. Stop and think about

4 Cairns, James L. (2006): "The Inventor's Pathfinder." New York, NY: iUniverse, Inc.

5 An excellent biography of Tesla can be found at: www.teslasociety.com/biography.htm

some subject for a second. You're mentally verbalizing what you're thinking about, aren't you? Now try to imagine forming a rational thought about something without mentally verbalizing it. You cannot do it.

Verbal thought tells the inventor what the limits are for his materials and manufacturing techniques, what his design goals are, and so on. It helps him avoid common errors. He can read about what others have done; learn what mistakes they have made. It sets many of his design's criteria and provides a means to communicate his work to others.

Mind's eye visualization, on the other hand, allows the inventor to mentally construct, alter, manipulate, destroy, rebuild, or operate the objects of his design completely within the space of his mind. Like Tesla, he can mature his invention to the point in which it is ready to build, without ever setting it down in a more tangible form.

Mind's eye visualization is the imagination's workspace; mental verbalization provides the discipline to keeps what's done there practical. Both are needed to create good, sound inventions.

In my own design endeavors, once I pick a problem to pursue I move it into my mind's eye workspace right away. I try to get as close to the design's final embodiment as I can within that space. On very rare occasions, nearly complete designs emerge in just a few minutes. Much more often, though, it takes months of catch-as-catch-can concentration to iron out details.

It's rare that mind's eye visualization is discussed as an academic engineering topic, even though it's obviously an extremely valuable asset. I believe that, like mathematics for instance, one can learn how to better use it through appropriate exercises.

Here are some exemplary ones that will give you a better idea of what I mean, and possibly help you improve the use of your own mind's eye. Please do not become discouraged if you find the exercises difficult at first. They should get easier with practice. Think of the exercises in the next few paragraphs as a sort of daydreaming with an agenda. Start simply. Close your eyes or stare off into space, whichever you find most comfortable. Relax. Try to clear your mind of everything else. It's very important that you're not distracted. Focus on the object of the exercise.

As a start, imagine a common, rectangular sheet of paper; let's say with the dimensions of typing paper. Picture it being red on the back-side and green on the front. In your mind's eye, picture it there in front of you. Then move it around, nearer and farther. Rotate it to different orientations. As it moves, follow the colored sides and keep track of where they are. Let the sheet just slowly tumble around in virtual space.

Then bring it to a stop before you with the green side facing you. Mentally fold it so the two diagonally opposed corners meet. Now see what it looks like. See where the green and red wind up. Repeat this frequently, folding the paper a different way each time, until it becomes easier for you. Remember, it's just a sheet of paper; you can do whatever you want with it.

In a second, slightly more difficult exercise, imagine an elongated, solid cylinder. Call up the vision in your mind's eye. Just let the cylinder slowly tumble around in space so that its orientation changes continually. Color one end of it blue and the other end yellow. Keep track of the colors as it moves. Make it slightly transparent. Now move inside of it, so that it's tumbling around you while you're fixed in space at the cylinder's geometrical center. Stay relaxed. Keep track of the colored ends as the cylinder moves around you. As you do this repeatedly it will become more comfortable.

Now try a third, even more difficult exercise. Let the cylinder become a hollow tube with closed ends. Imagine there's a ball within the tube. Let the tube slowly rock back and forth like a teeter-totter. Visualize the ball as it rolls from end to end in the rocking tube. Move inside of the tube and follow it from there. You don't have to dodge the ball. You're only there in spirit; let it roll right through you.

Here is a final example closer to home. It's one I like, and variations of which I often do myself. Imagine you are approaching the front door of your house. Picture the walkway, the door, the doorknob, and all of the exterior details of the house. Take your time, letting them all come into focus as you picture them one by one. Try not to miss anything. Reach out and take the doorknob in your hand; open the door, and go inside. Picture your surroundings just inside the door. When you have a clear mental vision of that space, walk about from room to room doing the same thing in each one, patiently visualizing as many of the details as you possibly can. When you find yourself in the dining room, rearrange the chairs.

By now you have the idea and can make up as many simpler or more difficult exercises as you wish. Or, you can find some other examples on the Internet[6]. Let them challenge your imagination's limits. Try to get comfortable in the virtual spaces you create.

Improving the use of your mind's eye will help you develop mental images of your inventions and better understand how things work. The ball in the tube, for instance, is just a simple assembly with two parts. By imagining it, you can experience how it functions. Moving inside it, you can get a better feel of how the parts interact.

6 One of many websites with exercises is: http://www.successconsciousness.com/index_000005.htm

The house is a more complex assembly. By rearranging the furniture you are creating a slightly different virtual reality; you are modifying the assembly. Entering mind's eye space advances your design capability well beyond what's possible by sketching on paper or on your computer. It allows you to pass into an imaginary world where you can create things, put them together, and take them apart at your leisure.

If you tried to visualize a very complicated device, like a washing machine for instance, you would probably find you couldn't visualize it all at once, just as you wouldn't see a whole house when you enter the front door. Instead, you'd visit its various spaces one at a time.

Continuing on with this general mode of thinking, here's the sort of virtual-space visualization that goes into product design. Keep in mind this is simply an example. It doesn't represent the design of an actual device. It just demonstrates the sort of thought processes typically involved. Take as the object a manual can opener of the simplest, common sort that has two hinged, plier-like handles. It clamps to the rim of a metal can. When you turn a T-shaped knob, a disc-like blade advances around the inside of the rim and cuts away the lid. Let's consider what the mental imagery might have been to design such a can opener. The opener has several cooperating parts that must all work in harmony to open the can. Here are the basic functions involved:

- Closing the can opener's handles simultaneously:
    - » clamps the opener to the can;
    - » forcibly engages a gear against the underside of the can's rim; and,
    - » causes the disc-shaped blade to penetrate the lid at the rim's inner edge.

- Next a T-shaped knob on the opposed end of the gear's axle is turned. Turning the knob advances both the blade and the gear along the rim, cutting out the can's top.

Now imagine a simple, cylindrical soup can with a rim around its lid. Visualize both the can opener and the can in your mind's eye. Try to clearly envision the whole scene. Zoom in on the can's rim and study its relationship with the can's side and top. See the can opener approaching the rim. Set the blade on the inner edge of the rim. Mentally squeeze the handles and watch the gear and blade act together to clamp the rim. Go right up there to the point where that's happening and study how the blade must be positioned to intercept the lid/rim junction. Look closely at the gear as it bites into the rim's lower edge. Put yourself right there. Study how the blade and gear act in unison. Next back out a bit and visualize how all the parts are positioned in the assembly and about what size they are. Start the assembly running. The T-shaped knob and gear turn slowly, and the blade cuts the lid away cleanly.

As you experience how the various elements of the assembly cooperate, you'll be able to mentally modify, add, or delete parts until you are pleased with the design.

While concentrating on the opener/can interaction, allow yourself to become various parts of the assembly. Start out by being the can's lid. See what you experience as the can's opening proceeds. Take other components. Be the gear for a while; then be the blade. Experience how that all goes. You can easily see how such visualization could help you develop the opener's design without ever setting pencil to paper or generating a computer model.

If you were actually designing a can opener, or other assembly, you wouldn't think of it as a 2-minute exercise, but instead as a project whose output is the virtual prototype of your invention's ideal embodiment. You'd probably revisit the imaginary can opener frequently over a period of weeks or even months before you were satisfied with the design. You'd make notes and simple sketches in a journal to help recall what you did. You'd be working like an inventor.

Practice exercises like those just described any time you have a quiet, relaxed interval and not much else on your mind. When you wake up at night and just before you get up in the morning are great times to concentrate. Start to live part-time in the virtual world of your inventions. Others will think you're strangely distracted. They'll be right.

Please go to the Quiz on page 259.

You should now be able to respond correctly to quiz items 1 through 9. The answers to these and all other quiz items are found in the ANSWERS Section on page 271.

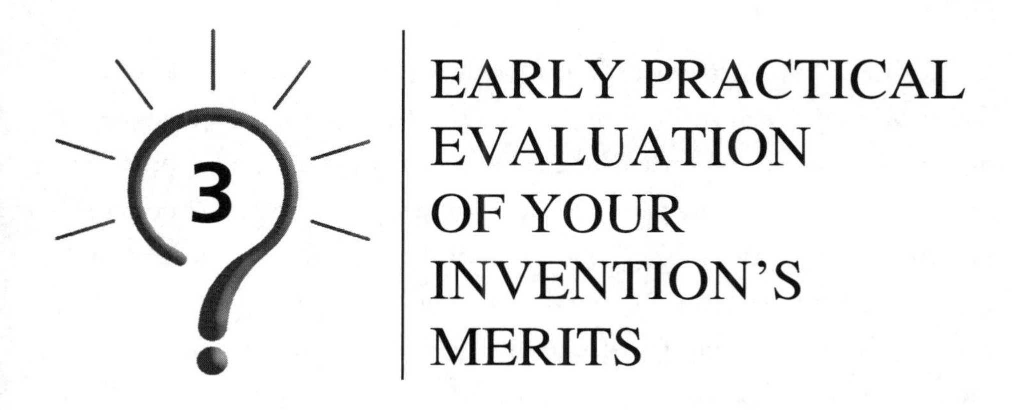

3 EARLY PRACTICAL EVALUATION OF YOUR INVENTION'S MERITS

Having taken a side trip to explore mind's eye visualization, it's now time to return to advancing your invention. By now you have chosen a problem to solve, and have found the best way you can think of to solve it. Here's how to move on.

First, advance your idea as far as you can within your mind's eye. Imagine you're tiny and like a spirit can travel over, around, and through your invention's every part. Become a voyager within it. Visit all its various elements. If it's an assembly, travel around and see how the components interact. Do they get along well together? Try to resolve any conflicts. As the vision matures, mentally assemble and disassemble it. If it has moving parts, turn it on and watch as it runs. Be sure not to miss any critical features. By any means possible, reduce its various elements to their most basic forms. Throw out unnecessary bits and simplify others. All non-functional details go. Ruthlessly seek simple, economical beauty. As you hone your design, remember that each component and feature, no matter how small, comes at some cost. Spartan designs are the winners.

When you have a crystal-clear image of your innovation and you're satisfied that it cannot be improved further by thought alone, take a comprehensive, in-depth look at what others have done. If in the end you find existing-product aspects that are superior to yours and they're in the "public domain," meaning that they're not private property, use them. If they're not in the public domain, try to design around them. Information in a published patent or patent application that is not claimed by the inventor is in the public domain, whereas the information that is covered by the patent's claims is not. In our later discussions of patents and patent searching, you'll see that's often easy to design around the claims of close-art patents. That might seem like cheating. It's not. If you can find a unique solution that's only slightly different from what's already been patented, you've made a contribution you can claim. Remember, Thomas Edison did not invent the incandescent light bulb; he's the one who made it work.[1]

With the idea for your invention firmly in mind, do a few more steps to test its merit. Don't forget, you're about to embark on a long, torturous journey with your idea, and you want to be confident that it's a traveling companion that deserves your unswerving fidelity. Here are some steps to measure its worth.

First, if your invention has more than one component, write a complete step-by-step assembly procedure. Include a packaging procedure as an appendix. Anyone who's ever put together a piece of IKEA furniture knows what this means. This exercise often reveals assembly difficulties not otherwise apparent and can uncover design problems solved more economically at an early

1 He claimed to have found thousands of ways not to make a light bulb before finding one that worked. See: http://www.bukisa.com/articles/108679_thousands-of-failures-but-thousands-of-patents-thomas-alva-edison

stage than at a later stage. Even highly trained engineers sometimes arrive at this point only to find that their devices are impossible to assemble. It's very embarrassing and, if not caught early, expensive.

Second, create a detailed parts list for your assembly, including packaging materials. Then try to determine what the individual parts will cost to produce in the lot sizes you expect to manufacture. Talk to a machinist, molder, or whatever specialist will eventually be needed to manufacture your parts. Ask for some rough estimates. Don't fully share your ideas with them yet. Just give them some exemplary piece-part sketches. Add up the cost of all the parts and multiply by six. That will give you a very rough idea of what it will have to sell for to yield a minimum profit. If you plan to sell it wholesale, try to estimate what markups will subsequently be added. That gives you a rough- idea of the over-the-counter price. Will the marketplace support that price?

Third, and this is important, prepare an excruciatingly detailed presentation describing your invention and its workings. Present it to at most one person, a trusted friend or family member. One of your parents will do. Give your presentation verbally, standing at a lectern whiteboard. Extemporaneously make whatever sketches you need to express your points. If there's no one to confidentially share your ideas with, just pretend you have an audience and lecture to the opposite wall of an empty room. That might sound a little weird, but you'll be surprised by what you'll learn. Something different occurs when you express your thoughts verbally and on your feet rather than sitting in silent concentration at your desk. It's rare that one fails to learn something useful this way. Please give it a try.

Of course, the lecture raises some real cautions. You have little or no ownership protection yet. So, if you choose your one-person audience unwisely, he could steal your idea or fail to respect your trust of confidentiality and blab your ideas all over town.(Sorry, even mom or dad could be a security risk here!) A more common dilemma is that your sounding board might proudly suggest fairly obvious extensions of your ideas and claim ownership of them. You would soon have thought of them anyhow, if you hadn't already. That's serious, and very awkward. You might have inadvertently and unwillingly acquired a co-inventor with some actual rights to your invention. But probably none of that will happen and you'll simply get some good input, mostly from yourself.

When your imagination has taken you as far as it can and your invention's ideal rendition is vividly in mind, you are ready to pass through the gate at [5] and continue your journey onward.

Commercial Evaluation of Your Invention's Merits

At this point you have a fairly good idea of how your invention will be made and a rough idea of what it will cost. Now, you need to make a judgment on whether or not it is likely to be commercially successful.

The chore of evaluating of your invention's commercial potential is an annoying but necessary bit of drudgery, not nearly as exciting as working on the invention itself. But you have to do it. You're about to put at stake a substantial amount of your time, your good will with family and friends (who frequently will not share your passion for sacrifice), your savings, and sometimes your health. You better be sure it's worth it. Although not at all precise, there are some ways to measure the commercial potential of the project you are about to take on.

Confidence in your invention's commercial viability can be gained (or lost) by considering the following points. Be as critical as you possibly can in your evaluations. You will not yet have very detailed information on some of the elements you need to consider, but do your best to answer the questions based on what you have.

Is there a demand for it?

Yours might be a great invention that serves a useful purpose, but one that few people need. Or, it could have very widespread applications, but no one really wants to use it. Either of these situations is fatal.

At some point before you invest too much, you'll need to judge your invention's potential market size to determine if the rewards will justify the effort. It's a wonderful feeling to invent something that solves a critical problem, but not very satisfying if no one wants to buy it. Remember, you're about to take a big risk. If you're successful, you should expect a correspondingly big reward.

Suppose you have improved a well-known product such as the can opener of our previous example. You know already that people need can openers. So, there's almost surely a demand for your improvement. Given that, you could fairly easily determine in a gross way the existing worldwide market for can openers. Then you could conservatively guess at the fraction of that market that you might capture with your improved opener. Your estimate will be crude, but good enough to see if the development is promising.

Suppose, on the other hand, you have invented an improvement to a device that's not very widely used. Then it's not so apparent whether or not there's a need or a strong public desire for your product. Consider the invention shown in Figure 2. It allows a weary gentleman to rest his head comfortably against the bathroom wall while urinating. There were already similar prior-art devices that performed the same function, allowing an extended hand to rest against the wall for support. The earlier invention was apparently not successful, because the patent covering it, U.S. Patent 6,061,842, was allowed to expire due to non-payment of maintenance fees.

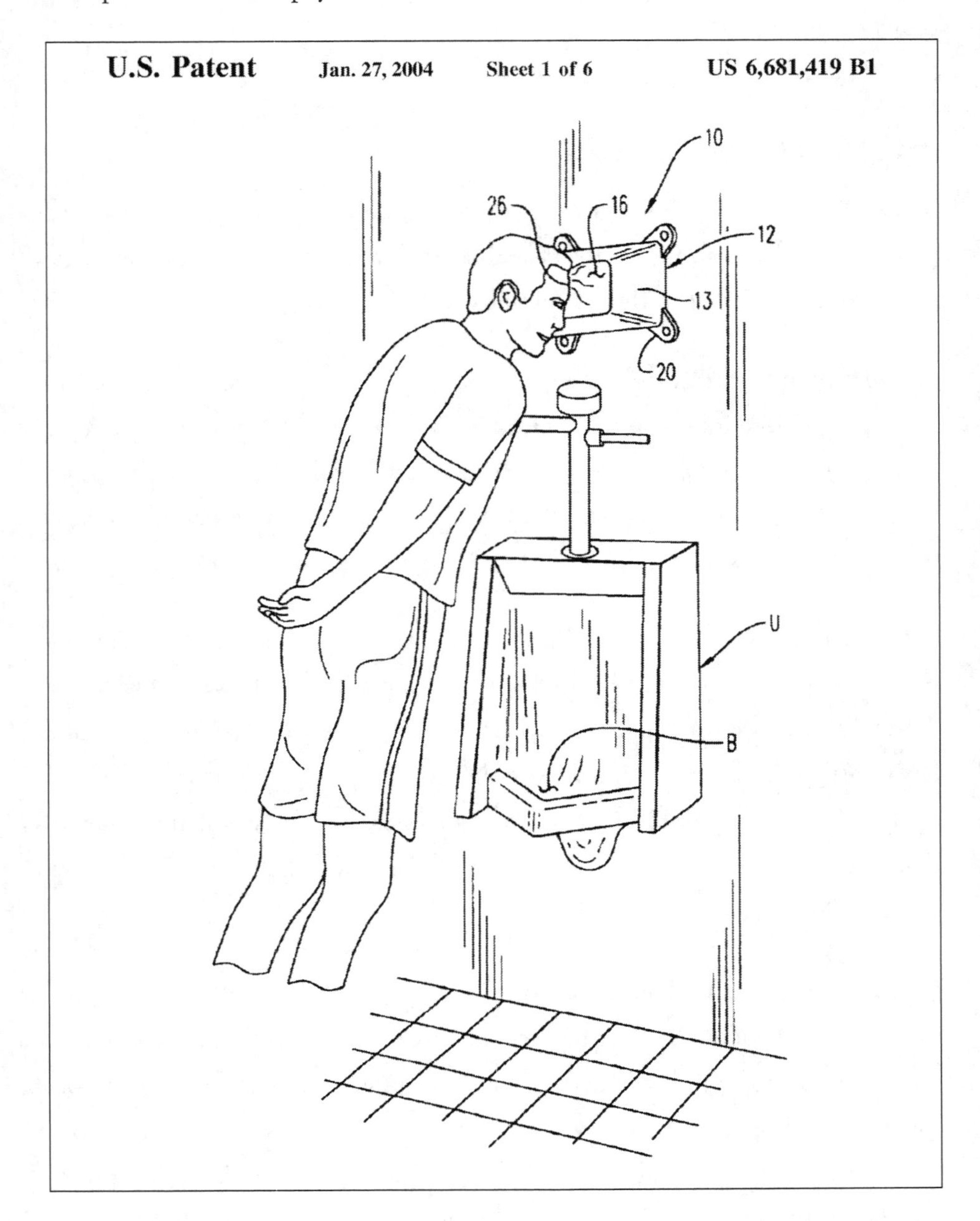

Figure 2

One might imagine that the inventor of the Figure 2 urinal headrest was just as exuberant about his device as every one of us should be about our own creations. And, as any inventor might have, when setting out to weigh its potential market he could have started by roughly establishing how many urinals there are in the world. He might then have taken a guess at what fraction of those urinals would be outfitted with the headrest, multiplied that by the unit price, and arrived at his projected revenues. Thinking expansively, for really up-scale men's rooms he might also have foreseen disposable headrest covers, and even dispensers for the covers. And the covers could have been printed with advertisements, bringing in another source of revenues. All in all, the overall sales projection would most likely have been somewhat staggering.[2] But, like that of its predecessor, the urinal headrest patent has expired due to non-payment of maintenance fees. What went wrong?

Apparently no one wanted to use it. Would you rest your face against a cushioned surface in a public men's room? Doesn't sound very appealing, does it? Probably in this case the inventor could have done some survey work to judge the public's opinion of using such a device before going forward with his patent. Similarly, he could have asked some hotel, gas station, and restaurant chains if they would consider installing the head rest. And, he could have looked at patents for similar inventions to see how they fared over time. In a later section of this book we'll see how easy that is to do. In short, he could have done a number of simple, inexpensive things to gauge whether or not there really was need and desire for his invention. The survey results probably would have been discouraging.

2 There's absolutely no reason to believe he actually went through such an evaluation; but this is just an example, so it doesn't matter.

There's a lesson in this example: Do not go forward until you have done as much research as you practically can to be sure that there's a demand for your invention.

That's obviously good advice, but it's more difficult to accomplish if your invention's a product that previously didn't exist at all. As an example, in 1950 University of California physicist Hugh Bradner invented the cellular neoprene wet suit.[3] It was the type now used by divers and surfers worldwide. Hugh never made any money from the wetsuit invention. He did not intend to seek any personal gain, granting his ownership in the technology entirely to the University. The University could have profited greatly from it, and didn't. It was observed by their patent committee that there were only a handful of SCUBA divers in the world then, and a market so small didn't merit the modest filing expense. No patent application was submitted, and a commercially important opportunity was missed. They were foiled by a common miscalculation: his was a completely new, enabling technology for which there was no recognizable market. They did not see the potential.

The few hearty souls then surfing and diving in California's sometimes 45° F coastal waters had no good way to keep warm. One might suppose that's why there were so few of them. In any case, a way to keep them warm was needed, but until then the technology didn't exist. Consequently, there was no *established* market for it. However, as we all know the wet suit soon created its own market. That's a fairly common situation that sometimes is not adequately considered when deciding whether or not to invest in an innovation's development. Here's the point: The more revolutionary your emergent technology is, the less likely it is that there's an existing market. That doesn't mean there's no potential. There was little market for lamp cord until Edison perfected the light bulb.

3 http://stokereport.com/rant/california-cool-how-wetsuit-became-surfers-second-skin

To summarize, in estimating whether or not your product is commercially worthwhile, try to evaluate the current market for similar products, if they exist, and the fraction of that market your product might eventually capture.

If no competitive products currently exist, you're stuck with following your own instincts. Before going ahead, convince yourself that your invention really does satisfy an unfulfilled commercial need. If your technology is completely new, it could take customers a long time to abandon present work-around solutions in favor of something untried. Be prepared for a long wait to get a return on break-through technology.

Evaluating the commercial potential of inventions is often at best very imprecise. Still, it's worthwhile to make the best estimate that existing data allow.

Will your product stand out from the competition?

Product recognition is important, particularly for consumer items. It's a real plus if your product has easily recognized attributes that make it stand out. As you design, try to highlight your product's unique features and functions. If you can, give it something that sets it apart from the rest.

Will you have enough protection to discourage "knock-off" competitive products?

To some extent this question involves the breadth of patent coverage you can expect to get. Some patents provide very little protection against others copying your product. Before you make a major investment of time and money, ascertain that you're likely to get sound patent claims on the unique features of your invention. You can gain that confidence through searching existing patents. In the later section dealing with patent searching, you'll see how to weigh your own proposed patent claims against those of existing, competitive products.

The importance of strong patent claims varies from invention to invention. If the life cycle of your new technology is very short and you have a good expectation to get an adequate return on your investment before someone else can market a knock-off version, you're probably OK with minimal, or even no patent protection. You might also be all right if you can quickly establish an unassailable market position or some other strong market-entry barrier to would-be competition.

Do you have the resources to develop it?

Here again, watch your optimism. It will be a lot harder than you think. It will also probably cost more and take longer than you expected; conservatively plan for a very long haul. Don't count on significant short-term revenues from the sale of products. Many inventions take four to five years before there are significant sales' revenues. Stretch your funds and allot your time accordingly. As we'll see in a later section, earlier returns to the inventor himself can be achieved by licensing or selling rights to use the invention, as opposed to waiting for revenues from the actual sale of manufactured products.

Can it be manufactured?

Surprisingly, you can design perfectly reasonable looking things that are impossible or horribly difficult to manufacture or assemble. If you don't have much manufacturing experience, try to get some expert advice without compromising your invention's secrecy. Try to iron out manufacturing difficulties early in the game. If they linger, they can cost you plenty downstream due to product returns, high assembly times, scrapped parts, and so on. Manufacturing and assembly problems are most easily eliminated by designing with production in mind. If your design does not minimize manufacturing difficulties, your product could wind up costing much more to produce than you anticipated, thereby pricing itself out of the market. I have experienced that unpleasant situation.

Many years ago I designed an electrical connector that could be plugged and unplugged underwater. It was supposed to be a low-cost and high-volume alternative to the more expensive competition already on the market. It turned out to be extremely difficult to manufacture, requiring very skilled assemblers, and even then the scrap rates were unacceptably high. In the end it was neither low-cost nor high-volume. We wound up with a huge inventory of products for which every sale lost money. Eventually, all of the connectors were scrapped. It was more economical to throw them away than to sell them. I had made the mistake of rushing to make tooling and parts before completely working the kinks out of the design. It was an expensive lesson.

The message here is: keep your designs absolutely as simple as possible. Do not commit to high-priced production tooling until you're sure all the bugs are worked out of your design. As noted earlier, clean, Spartan, well-honed designs are the winners.

Will it be user friendly?

It is one thing to design a device you can use yourself; it's another to design one that works easily and reliably in untrained hands. If using your invention is easy and straightforward, it's more likely to be used properly and work well. When people enjoy using your product, they will probably buy it again, and recommend it to their friends. Conversely, frustrated, dissatisfied customers will generate negative press. Once your product is on the market, you will be amazed at the careless handling and other abuse it will suffer at the hands of its users. Design it to be as customer-proof as possible.

Is it affordable?

Your product must be price competitive; otherwise, customers with a more economical alternative won't buy it. As an example from my own field of subsea power and communication hardware, it would be possible to design a wonderfully reliable $1,000 connector to be used on a $250 underwater camera. It could be the best purpose-built connector ever made, working reliably every time. But no one would buy it; they'd buy the $40 alternative that worked most of the time.

Will its market be enduring?

Consider what the life cycle of your product will be. You can get some idea of that by observing how quickly technology renews in your invention's particular field. For high tech electronic items such as computers, cell phones, and so on, the technology lifetime could be measured in months. Conversely, for the can openers of our earlier example the life cycle might be decades or even more. Sooner or later, though, your technology will be replaced no matter what it is. Can you expect a good investment return in that time frame?

Are there serious barriers to market entry?

These could include such things as health, safety or legal issues, industry standards, required permits, approvals, licenses, overbearing competition, etc. User reluctance to accept new innovations can also present a serious "soft" barrier that might melt away with time.

Unforeseen laws or regulations might block your product from being sold everywhere. For instance, some U.S. technology cannot be sold in certain foreign countries. The U.S. State Department International Traffic in Arms Regulations (ITAR)[4], and the Bureau of Industry and Security's Export Administration Regulations (EAR)[5] impose rather broad restrictions on the export of technology. Take any barriers you can identify into consideration when defining your potential market space, particularly if yours is a high-tech invention.

Another barrier to market entry might simply be customer refusal to adopt your invention no matter how good it is, even if there's no competition. That's a hard one to predict. Here's an example.

The invention pictured in Figure 3 on the following page, is a very clever, inexpensive safety device. It would be used in the case of a high-rise building fire. Most fatalities in such fires occur due to smoke inhalation. The device allows someone trapped in a hotel room or office to access breathable air by way of the toilet. Every toilet has an upward extension of the drainpipe that passes out through the roof. The user inserts a flexible tube through the water in the toilet bowl and up into the trap beyond the water level. He then has an unlimited supply of smoke-free air. By all reasoning this truly seems to be a fundamentally good idea. But, it was not commercially successful. Its inventor's attempts to place it with major hotel chains failed. They simply did not want to have such a reminder of potential problems in plain sight. (This would be a good trick to remember if you're ever caught in a building fire. Smashing a hole into the toilet at the right spot would also get you to the air.)

4 ITAR regulations can be found at: http://pmddtc.state.gov/regulations_laws/itar_official.html

5 EAR regulations can be found at: http://www.ntis.gov/products/export-regs.aspx

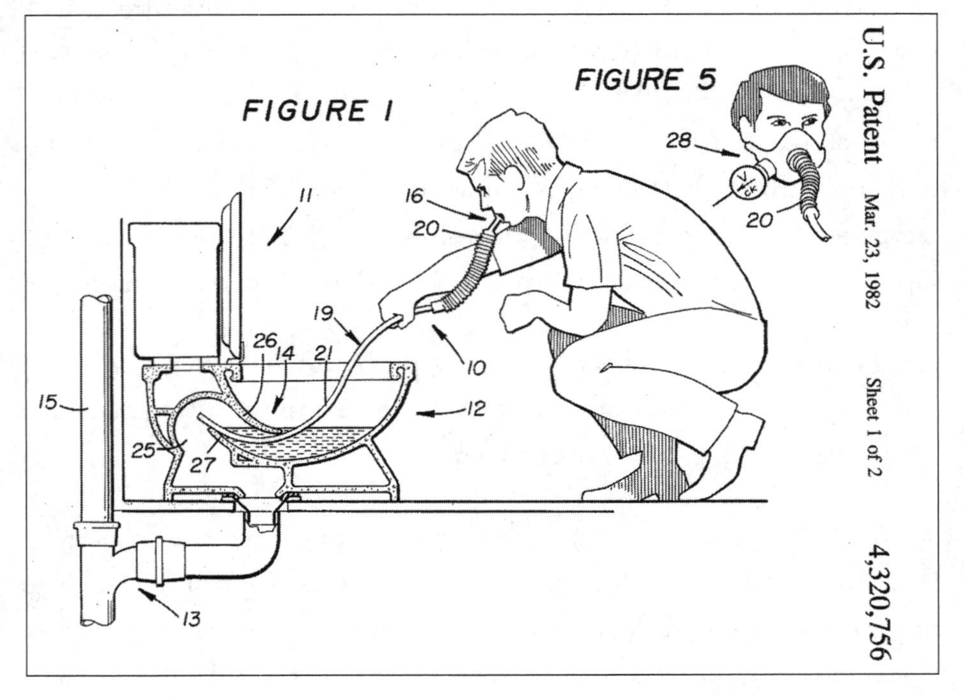

Figure 3

Are there any legally required tests that must be performed prior to your product's sale?

Find out what they are and design to pass them. An example of such tests would be those performed by Underwriters Laboratories, Inc. (UL) to obtain approval for household appliances[6] and those intended to comply with the guidelines of The Occupational Safety and Health Administration (OSHA)[7]. Such obligatory tests can be very expensive and time consuming. Anticipate them in your budget, and plan for them in your schedule.

6 Information is available at www.ul.com.

7 OSHA guidelines are found at: www.osha.gov/

Does it have a simple beauty?

This question might seem out of place, but designs that have a clean, elegant simplicity are more likely to be accepted and to work than those that do not. And, they're a pleasure to see.

If your evaluation of all the above questions is positive, it looks like you have something worth taking further. When you've arrived at this point, make drawings detailed enough to capture all your design's salient features.

If possible, you should also make some inexpensive models to see if your idea's principal features actually will work. Try to make them with your own hands if you can. That, too, is a great learning opportunity. You're not finished with your design work yet, but it's time to temporarily put it on the back burner. You'll return to it later. You've completed an important step of the inventive process: You have identified a "fundamentally good idea" that's worth pursuing.

It might occur to you at this early stage to seek a partner or try to sell some interest in your invention. That's probably a bad move. You really don't have much to sell yet, no matter how good your idea is. And even you, the inventor, won't know exactly what you own until you have an issued patent. Additionally, the earlier in the process you seek partners, the more ownership and control you'll concede for a given return. It's better to go as far as you can with your own resources. The next few logical steps along the pathway advance your invention's development with relatively little additional expense. Taking them will increase the invention's value, and might even allow you to get some funding without diminishing your ownership.

Please go to the Quiz on page 259.

You should now be able to respond correctly to quiz items 10 through 18.

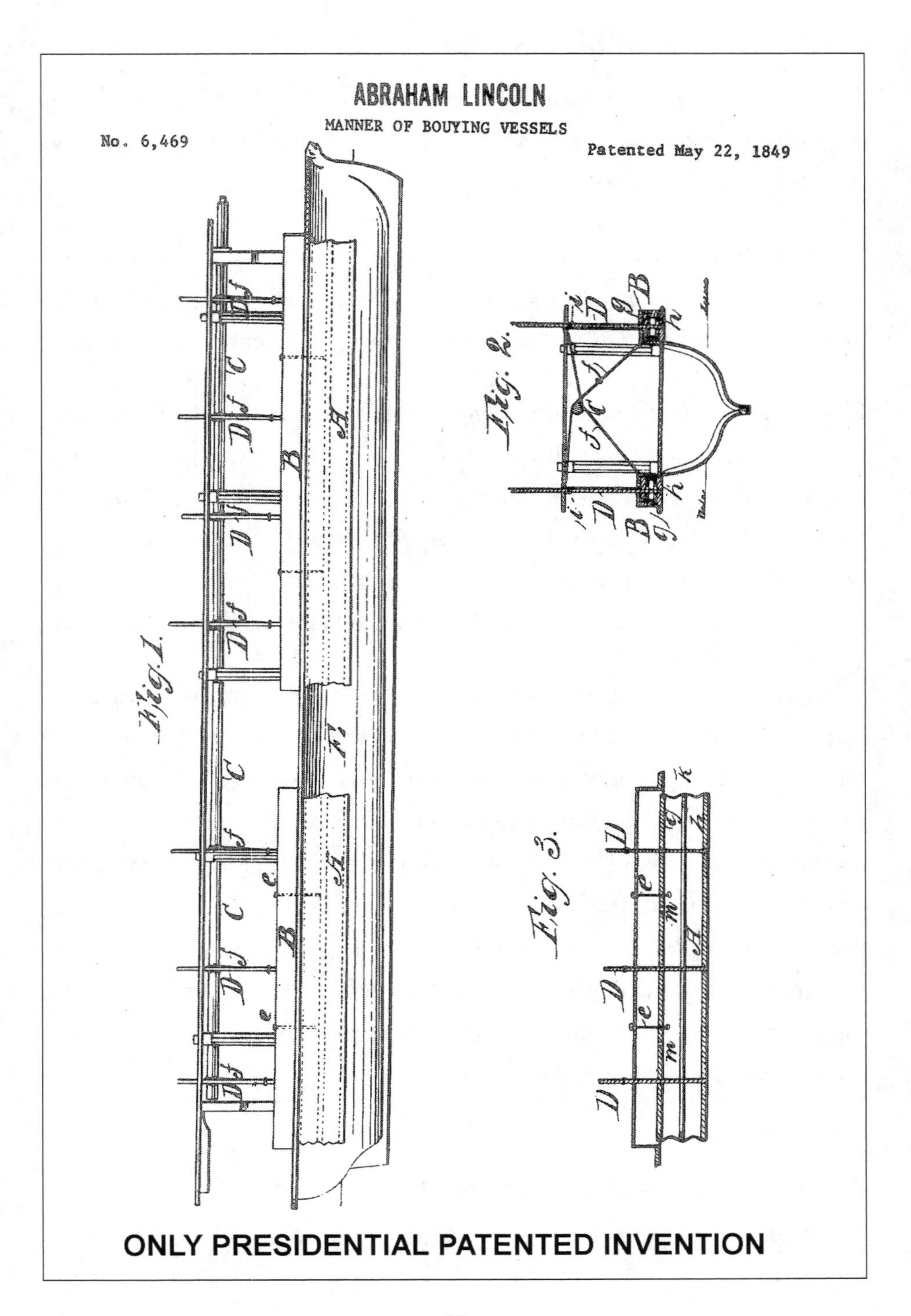

ONLY PRESIDENTIAL PATENTED INVENTION

BUSINESS SUPPORT & POSSIBLE FUNDING

By this point you have identified your invention, weighed its practicality, and judged its commercial potential. It's time now to take another short side-trip to Map point [6] to look at some of the preliminary business aspects of going forward.

In a later section we will discuss in detail various ways to cash in on your ideas. Stated briefly, you will learn that the most common ways to cash in are to:

- Sell outright all or part of your interest in your invention;
- Grant another entity the rights to use your invention for some period of time; or,
- Start a business to manufacture and sell your own products.

Whether your eventual goal is to permit someone else to use your technology or to manufacture the product yourself, you should have a business plan. Taking the time up front to formulate a good plan is one of the best investments you can make. The plan should describe how you intend to deal with *every* aspect of your new venture, not just the financial aspects. You do not have to blindly follow the plan. The plan can change from day to day, if necessary. But, it at least documents a thoughtful approach to your project, and to the changes made to that approach as the project matures.

Starting off without a plan is like beginning a trip without knowing where you're going, or how to get there. If you cannot put into writing how you expect your efforts to evolve, you're not ready to begin. It simply does not make sense to reduce your chances of success by not having a thoughtful plan.[1] Try to develop a draft of the plan yourself; then get some experienced help to flush it out. As you'll see, you can probably get all the help you need right in your own hometown. You will likely be surprised by the great, inexpensive assistance available locally. Much of it is sponsored by the U.S. Government, which is very supportive of entrepreneurs, recognizing that they are the ones planting the seeds for future growth of the Nation's economy. It's a fact that small businesses produce 13 to 14 times more patents per employee than large companies.[2] The U.S. Small Business Administration (SBA) has Small Business Development Centers (SBDC)'s in all 50 states, Puerto Rico, Washington, D.C., and in all U.S. territories. You can find the location of the SBDC office nearest to you on their website.[3] The Centers provide a wide variety of services, including helping you formulate your business plan. They can show you how to develop financial plans, and help you with organization, engineering, technical issues, marketing, help organize patent protection, and many other elements of your business. The services are completely free, and available to anyone wanting to start a new business or upgrade an existing small busi-

1 According to Scott Andrew Shane in his 2008 book *The Illusions of Entrepreneurship: The Costly Myths That Entrepreneurs, Investors, and Policy Makers Live By* about 50% of all new businesses fail in the first five years due to poor planning or insufficient resources.

2 www.stopfakes.gov Training module "Importance of IP Protection to Small Business."

3 For locations and other information go to: http://www.asbdc-us.org/.

ness. The centers are staffed by professionals, and augmented by qualified volunteers. They are there to encourage you, and to help you keep from making serious mistakes. They want you to succeed. That's their mission. In addition to the abovementioned free services, the SBDC's frequently offer workshops and other forms of training at very modest cost.

You are about to embark on a new commercial activity, and even if you think all your ducks are already in a row, go visit the SBDC nearest you. When you have written a draft of your business plan, take it to them to help you polish it. You'll be glad you did.

Before discussing your invention in any detail with any of the herein mentioned support groups, or anyone else for that matter, it is wise to obtain some ownership protection for your ideas. You can do that by filing a low-cost provisional U.S. Patent application. In the next section of this book you will learn how.

Business Incubators (BI's) provide another means to get very competent, valuable support. They promote regional economic development by providing entrepreneurial companies with an array of business support resources and services.

Although there's a lot of overlap in the services offered by BI and SBDC centers, there are also a lot of differences. One difference is that for the most part the BI's are not free. Some provide their services in return for an equity share in the businesses they support. Others charge a modest fee. In either case, the BI client can have an office, an address, receptionist, copier, phone, facsimile, internet access, conference facilities, technical, business and often machine shop support, and just about everything else needed to conduct a nascent business.

You can find the location of BI's in your area on The National Business Incubator Association's (NBIA) website.[4] The site also contains valuable advice on how to choose an incubator that's right for you. There can be wide differences from one BI to another, and so it's important to choose one for which your business qualifies and which will fulfill your needs.

Studies have shown that successful completion of a BI program greatly increases the likelihood that a start-up will stay in business for the long term. Nearly 90 percent of BI graduated businesses remain in business at least three years after completing the program.[5] That statistic reflects the fact that unlike the government-funded SBDC's, which are free and must accept anyone needing their help, many BI's thoroughly vet their applicants; only the most promising are selected for participation[6]. At some BI's applicants must have a promising business concept and a practical, well developed, business plan already in place to be considered for participation.

It's easy to see why the Small Business Development Centers and Business Incubators frequently work hand-in-hand. Often the SBDC's act as a preparatory step in advancing a business concept to the point where it becomes a candidate to enter an incubator program. Conversely, if one first applies to an incubator, and is not sufficiently prepared to be accepted, he might be advised to go to a SBDC to come up to speed before reapplying for the incubator program. Both of these organizations are wonderful resources, and there's probably one located very near to you.

SCORE, a U.S. Small Business Administration (SBA)[7] partner, is another extremely useful resource for start-up businesses. It's a non-profit organization

4 For the location of incubators go to: http://www.nbia.org/links_to_member_incubators/index.php

5 See the NBIA website section for entrepreneurs: http://www.nbia.org/for_entrepreneurs/

6 Business incubators are different around the country. Some carefully pre-qualify their clients; others do not.

7 Sorry about all of the acronyms. Some of these programs and agencies are better known by their acronyms than by their names.

that offers small-business advice and training as a public service. Its volunteers are both active and retired businesspeople with a wide variety of expertise. Their advice is free and confidential. SCORE offers both prepared guidelines and one-on-one mentoring. If you're a beginning entrepreneur, you should contact a SCORE office in your area for help. To find out more about SCORE, including the location of your nearest SCORE office go to: **www.score.org.**

Early Funding:

One of the important ways that both SBDC and BI counselors can help is in obtaining early-stage financing through a number of different avenues. When you seek financing, and in fact when you seek most any sort of help, including participation in a business incubator program, you will most likely have to form some sort of business entity: a corporation, for instance. That is easy and inexpensive to do, and the BI or SBDC will help you do it.

Once you have created a formal business entity and filed a provisional patent application on your invention, you're ready to look into financing.

Of the various ways start-up funding might be obtained, U.S. government programs are probably the least burdensome. For the most part, they're as favorable to the inventor as the law allows. They are there to help. They also have an important aspect most other forms of financing do not have: they do not require you to dilute the ownership of your business. They are not looking for equity in the companies they support.

The SBA has Small Business Innovation Research (SBIR) grant programs to stimulate development of a broad range of critical technologies. These programs are funded through 11 government agencies such as the Department of Defense (DOD), the Department of Energy (DOE), the Department of Agriculture (DOA), and so on. The only qualifications needed to apply for

SBIR grants are that: The entity receiving the grant must be American owned; the principal researcher must be employed by the entity; the recipient must be a for-profit entity, and the entity must have fewer than 500 employees.

Proposals received by the agencies making SBIR grant awards are judged upon their various merits. Small businesses chosen to receive the awards then begin a three-stage program.

The first is start-up Phase (I) to prove the technical merit or feasibility of the proposed new technology. Phase (I) funding is up to $150,000, and lasts about 6 months.

Phase (II) is a continuation of Phase (I) plus positioning the innovation for commercial release. The grant to enter Phase (II) requires that Phase (I) has been successfully completed. Phase (II) can last up to two years with a total award not exceeding $1,000,000. Under some circumstances it is possible to receive a second Phase (II) grant to further the Phase (II) work.

In Phase (III) the innovation moves into the marketplace, and the innovator receives no more SBIR funding. However, it is very common for the funding agency to advance additional R&D funding to adapt or improve the technology for the agency's specific use. It is the intent of Congress that agencies support the SBIR awardees' commercialization of the funded work through Phase III contracts. So not only does the SBIR program fund the development of your technology, it gives you an immediate market for your products.

SBIR grants permit you to pay yourself and your employees a reasonable salary plus all other direct costs incurred in the development of the covered technology up to the award limit. Funding agencies also have the latitude to exceed the award limits by up to 50 percent in special cases.

Unless otherwise required, the government doesn't publicly disclose proprietary data of SBIR-supported inventions for four years after the last award phase was completed. That courtesy gives the inventor a good head start at establishing his market position.

To the extent permitted, all data rights in the SBIR funded technology remain with the award recipient. The government does retain "march in" rights which allows them to assign licenses to third parties in the case they determine that the funded technology is not being made reasonably available to the public.[8] Those rights are practically never exercised.

If your work is done in cooperation with a non-profit university or federally funded R&D facility you might be eligible to participate in another government program similar to that of the SBIR. The SBA also administers a Small Business Technology Transfer (STTR) program intended to help advance innovations wherein the small business is in partnership with a non-profit research institution.

Government programs such as SBIR and STTR are intended to encourage and nurture creativity. That means they are there for you, the entrepreneur. If you can get either SBIR or STTR funding, they are a good deal no matter where you are along your invention's development pathway. And the earlier you can receive the funds, the better off you are.

In addition to grant programs, there are SBA loan programs. The SBA generally doesn't actually loan money; instead, it acts as guarantor for loans provided by private institutions such as banks. The SBA's guarantee reduces the lender's risk, thereby making it easier for the small-business owner to acquire reasonable debt financing. You will probably be asked to sign a personal guarantee to pay back an SBA loan in case your company defaults on the payments. Be

8 The 1980 Bayh-Dole Act [35 USC 200-212]. Details at: http://www.autm.net/Bayh_Dole_Act.htm

careful in doing so. If you have other equity holders in your business keep in mind they will all directly or indirectly benefit from the loan, but if your business cannot make the payments, you, as the founder or director will probably be the one taking on the whole repayment burden. You can certainly understand why a bank would want such a personal guarantee, but I have never taken an SBA loan for that reason.

Applying for any of the abovementioned government-backed financing opportunities can be intimidating to a first time applicant. But your local BI or SBDC center should be able to lead you through the process. Here, again I urge you to take advantage of their help.

Remember, we're still not far beyond the gate at [5] on our Map. As we entered this part of the path I mentioned that it is not advisable to sell interest in your invention at this early stage because beyond this point its value should begin to increase remarkably with little financial outlay on your part. However, if you cannot qualify for any of the government programs just described, and you cannot go forward without taking in additional capital, your local SBI or SBDC center might be able to line you up with equity-seeking investors.

Generally the SBDC personnel are thoroughly in touch with the economic landscape of your community. Drawing upon their networking experience they can help put you in touch with potential investors, and facilitate your meetings with them. If there is local investment money available, they'll help you find it.[9]

The BI's will do the same thing, but only for their clients. Additionally, they organize workshops in which their clients present their innovations to groups

9 Some SBDC personnel might be more eager to help you line up SBA loans than finding private investors. Being SBA organizations themselves, they are more familiar with that product. Look at all options available to you before accepting funds.

of venture capitalists and angel investors. The BI counselors help with presentation skills, so that their entrepreneurs can effectively pitch their proposals to the potential investors. They help polish the presentations, and provide the workshop venue. Successful, seasoned BI volunteers might review the presentations prior to the event. In short, the BI does everything to present the new technology and its creator in the best light to prospective investors.

One other way to attract investment capital is to ask family and close friends to buy some equity in your start-up business. Although I have had SBIR funding from time to time in my business ventures, my favorite and most frequently used method of raising capital is through family and friends. It's a great feeling to have those closest to you participate in your venture. They share in your triumphs and woes. And having their money at risk adds a bit of pressure. Mom & dad are going to ask frequently how their investment's coming. It's different than working with a stranger's funds. Even though you will still probably be trading away some of your equity at a bargain basement rate, at least the beneficiaries will be family and friends.

Aside from SBIR or STTR awards, no matter how an inventor obtains outside funds to further his efforts, he'll have to give up something: either autonomy or equity, or both. If early funding is acquired through private sources, the cost in terms of ownership and/or control loss will probably be very much higher at this early stage than it would be a little further down the road. My advice is to hold out as long as you practically can before diluting your equity.

Please go to the Quiz on page 259.

You should now be able to respond correctly to quiz items 19 through 28.

C. F. WASHBURN.

BARBED FENCE WIRE.

No. 249,212. Patented Nov. 8, 1881.

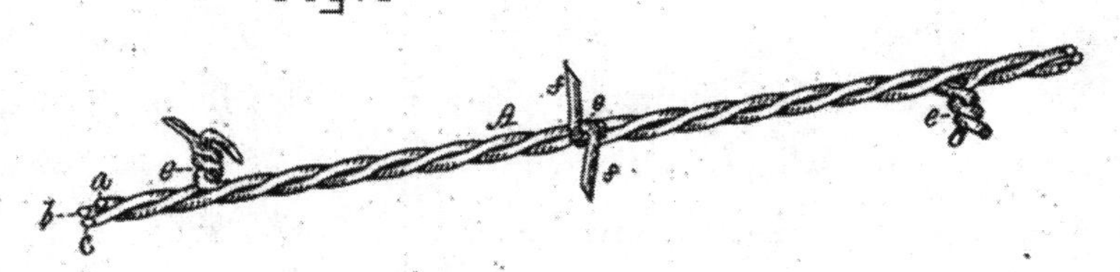

WITNESSES:

Julius Wilcke

N. H. Sherburne

INVENTOR:

Charles F. Washburn

By Gridley & Sherburne

Attys

5 INTELLECTUAL PROPERTY INTRODUCTION

Another side trip, this time to Map segment [7]. Taking these side trips from the main inventing route to learn what you're doing can be a real nuisance, I suppose. But it's better to be informed than to forge ahead and possibly make grave mistakes. The topics in this section get us off into legal territory.

Before travelling through the gate at point [8] you should spend some time learning how to claim and protect the ownership of your inventions. The term "intellectual property" encompasses creative innovations such as inventions, writings, music, art, and other products of the mind. It is defined and protected by contracts, trademarks, copyrights, trade secrets, and several types of patents. The inventor need not be an expert in these matters but must have a good general awareness of them to properly obtain and defend his rights.

This side trip on path segment passing [7] is a long one. The discussion begins with a sequential overview of *patents*, *trademarks*, *copyrights*, and *trade secrets*. They are all relevant to the legal protection of your creative work. After a brief introduction to these topics there are portions of our journey that will take you out of this book and on to various Internet websites. There you will find very comprehensive information on the each of the subjects, including well-prepared lectures and quizzes. If you study the material of this side trip well, you will be armed with a very good working knowledge of intellectual property protection.

Intellectual property (IP) laws vary throughout the world, so what's covered here does not necessarily apply to countries other than the United States. Be advised that even U.S. rules change frequently, and although this is current at the time it's written, it's a good idea to check for updates as you proceed with your own project. Even as this book is being prepared for publication, both U.S. and foreign patent regulations are in a state of flux.

You might have a world-class invention, but if you cannot claim and protect its ownership, it's probably of little value to you. Even if your invention is completely unique and otherwise patentable, its ownership might be compromised in several different ways. For instance, many employees are under some sort of actual or implied intellectual property agreement as a condition of their employment. The rules governing employment agreements vary from state to state. Typically these agreements require that an employee disclose any inventions to his employer. The employer can then determine if its content is within the scope of the employee's contract. If it is, but the company has no interest in pursuing it, the employer might give the inventor a release. If it's not, the inventor usually has an automatic release. If inventing things is part of what you were hired to do, your employer may have implicit rights to your inventions. Carefully check your job description and any contracts you have with your employer. If there is any doubt about your contractual covenants, check your position out with an attorney before investing much time and/or money pursuing your innovation.

If you're inventing things on your own, it's wise not to use any of your company's resources. If you do, the company will likely be able to acquire what are called "shop rights," which might entitle it to some fractional ownership.

Do your best to determine whether or not you have some obligation to your employer. If you do, try to get a release. Until there's some ownership determination, be careful in discussing your ideas even with your employer, as you have no protection yet.

Another way your ownership might be incomplete is if you have a co-inventor. Ask yourself whether anyone else has contributed materially to the invention. Were any of its features or functions suggested or developed by someone else? If so, the joint inventor is entitled to joint ownership. It's not enough to have been present or involved in the innovation's development or to have acted as a "sounding board" for the inventor. However, if the "sounding board" suggests even an obvious extension of your idea that becomes part of the patent, he might feel, and actually be, entitled to joint ownership. That's a very awkward situation. It's why I mentioned earlier that when you choose someone to review your designs, be careful. The reviewer might suggest an obvious modification, and become your unwanted co-inventor.

A legitimate co-inventor must have made a substantial intellectual contribution to the invention that results in at least one of the issued patent's claims to uniqueness. As an example, suppose a graduate student is asked by his professor to periodically introduce a measured amount of catalyst into a solution to maintain a constant chemical reaction rate. The student gets tired of doing that manually, as it has always been done before. He goes off on his own and develops a unique injection pump that senses the reaction rate and automatically introduces the correct amount of catalyst at the right times. It's the student who is the inventor. The professor presented the problem, but he contributed nothing to its solution and therefore is not a legitimate co-inventor. Similarly, if a laboratory classmate helped the student build the pump but added nothing intellectually to the device's concept, the classmate could not be legitimately considered a co-inventor.

On the other hand, if the student inventor who developed the injection pump used university facilities, the university could claim shop rights, and thereby possibly acquire fractional ownership.

In another instance, suppose the supervisor of a tire store wishes out loud that someone could find a better way to dismount tires for replacement. One hourly employee goes home and on his own invents a far superior dismounting mechanism. He has used no company resources in its development and has no formal or implied employee agreement. He files for and receives a patent on his invention. The supervisor might complain that he himself should be a co-inventor because he highlighted the problem. But even though the supervisor proposed the problem, he conceived no part of its solution. So, he's not an eligible co-inventor. The tire store might try to acquire shop rights to the invention, but it's not likely that they would be successful, as it was conceived and developed entirely without their resources.

Providing financial or other material support in the invention's development also does not qualify the contributor to be a co-inventor. If someone does help you financially, be careful to let him know up front that he cannot be considered a co-inventor on that basis. To grant such funds, however, he might insist on a contractual agreement giving him an ownership stake. If you accept his funds under those circumstances, he would contractually become the intellectual property's co-owner but not a co-inventor.

If you do have a legitimate co-inventor or if some percentage of your ownership is contractually assigned to a non-co-inventor, as you might have done in return for up-front development funds, for example, the assignee and/or co-inventor would have as much rights as you to exercise the resulting patent. No matter how small his part interest, if there's no contract stating otherwise, any patent stakeholder has full rights to exercise the patent. He may sell his interest or any part of it or grant licenses to others without regard to you or any other part owners. It is accordingly dangerous to assign a fractional interest without a definite up-front contract between the parties that clearly delineates the full extent of their respective rights and obligations to each other. It's advisable to formally make such a contract as early as possible. It's much easier to reach an amicable agreement before any significant money is in play or any attractive commercial opportunities arise. A partner's greed could quickly turn your inventing experience sour.

Still another way your ownership can be jeopardized is through the loss of your invention's patentability. You can give up ownership inadvertently by publicly disclosing your invention prior to filing a patent application. Any public use, sale, or open publication of the invention anywhere in the world more than 12 months prior to filing will prohibit the granting of a U.S. patent. Foreign patent laws in this regard are at least as restrictive as U.S. laws, so be extremely careful.[1] Even though it is your right to do so, it's wise *not* to openly disclose your invention either verbally or in print in any forum prior to filing at least a provisional patent application. If you must disclose, do so only with the greatest care. Anyone to whom you disclose details of your invention prior to the actual filing of a patent application should acknowledge in writing, that he is receiving proprietary, confidential information, and that he will not disclose it to anyone else. If you do not protect the confidentiality of your

1 More information about disclosure rules can be found in the Appendix I summary of the America Invents Act.

invention prior to filing, you run the risk of having someone else try to claim it by filing a patent application before you do.[2]

Sometimes it's awkward to have others agree to confidentiality, particularly if you're presenting your invention to a group. But you must ask them to do so. Even with non-disclosure agreements in place, if you disclose your invention to others for business or other reasons, be very careful what you say. Limit the information to the minimum necessary to accomplish your purposes.

Prior to your meeting, prepare a sign-up sheet headed by a notice that the signatories are about to receive your proprietary information and they agree to keep it confidential. Make sure everyone present signs the sheet. If they won't sign it, don't disclose.

Try to accomplish the objectives of your meeting verbally. If you do pass out printed material, recover it all at the presentation's end. Anything other than verbal communications should bear a statutory notice stating that the information is your sole proprietary property and should be held in confidence. Below is an example of such a notice to that effect. You should check with your attorney to ensure that the wording of the notice you use is suited to your own circumstances.

> *The information contained or disclosed by this document is the confidential and proprietary information of (owner's name), and all rights therein are expressly reserved. By accepting this information, the recipient agrees that the information and material contained therein is held in confidence and in trust and will not be used, copied, or reproduced in whole or in part, nor will its contents be revealed in any manner to others except to meet the specific purpose for which it was delivered.*

2 More information about this risk can be found in the Appendix I summary of the America Invents Act.

If you use slides in your presentation, they, too, should have at least an abbreviated confidentiality notice on them.

It's always best to resist the temptation to disclose before having adequate patent protection unless for some reason it's absolutely necessary. You can pretty much count on someone failing to honor a confidentially agreement in one-way or another. In spite of signed documents, when that happens you're in an uncomfortable spot; there's not much you can do about it once the cat is out of the bag. You don't have to worry about revealing details of your invention to your U.S. Patent Office registered attorney or patent agent, however. They are bound to keep your information secret.

The foregoing disclosure arguments have two main aspects: (1) You need to safeguard your invention against theft until you have established ownership through a patent or other means, and (2) you need to protect your patent rights by not violating any public disclosure rules prior to filing.

Any potential ownership issues should be resolved before going further. Once they are resolved and you're satisfied with your position, it's time to formally establish and protect your claim to your creative, original work. The next section discusses the various options for accomplishing that. It deals with *patents,* and *trademarks* issued by the United States Patent and Trademark Office (USPTO), with *copyrights* registered by the Library of Congress Copyright Office, and with *trade secrets.*

Please go to the Quiz on page 259.

You should now be able to respond correctly to quiz items 29 through 37.

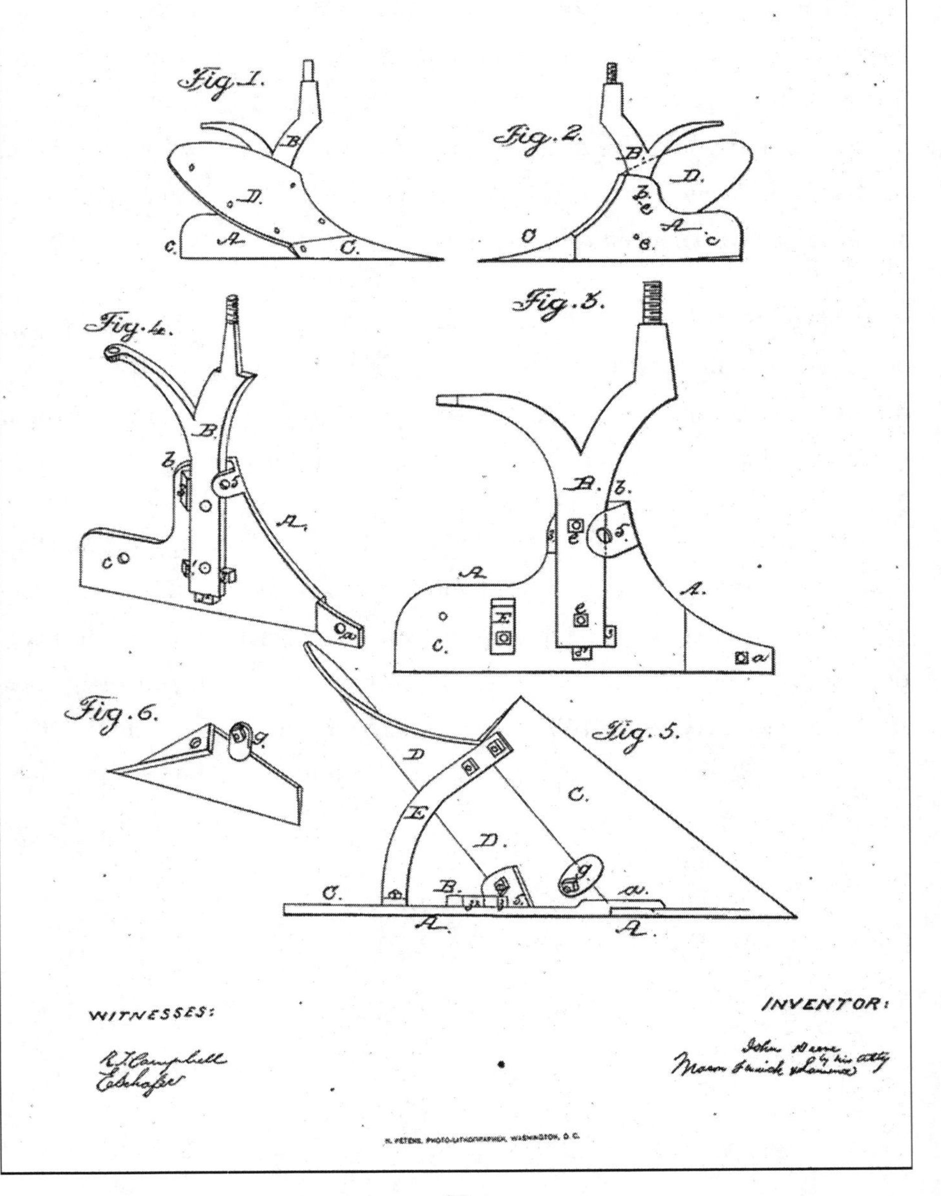
J. DEERE.
Plow.
No. 46,454.
Patented Feb. 21, 1865.
Fig. 1.
Fig. 2.
Fig. 3.
Fig. 4.
Fig. 5.
Fig. 6.
WITNESSES:
INVENTOR:

INTRODUCTION TO PATENTS

A U.S. patent[1] gives you the right to exclude others from making, using, offering for sale, or selling your invention in the United States or importing the invention into the United States. A patent defines your invention, and provides a tool for defending it. It doesn't defend it for you, however; that's your task. You must take the initiative to identify and pursue those who infringe on your rights.

Having a patent doesn't grant you any right to make or sell your invention. You could possibly get a patent on an improvement to a device that's already someone else's property, perhaps a device that's even patented. As an example, let's say someone has patented a unique wrench. You see it could be improved by adding a fold-out T-handle to obtain greater torque, and you successfully

1 For accuracy some phrases in this section have been copied or paraphrased from www.uspto.gov, the U.S. Patent and Trademark Office official website.

obtain a patent on your improvement. You couldn't manufacture your superior device without infringing on the wrench's basic patent. Your patent gives you no right to usurp someone else's ownership. On the other hand, the basic wrench's patent holder couldn't sell a product incorporating your improvement without your permission, as that would infringe on your patent's claims.

As another example, if you were granted a patent on an invention whose use would violate some law or safety regulation, you couldn't legally sell it.

The ability to enforce your patent can help protect what you've created. Your patent identifies you as the inventor, and its claims delineate the invention's scope. An analogy is often made between intellectual property and real property in which a patent is equated to a land deed and its claims equated to the land's boundaries. That's not a bad metaphor. A patent is a deed to the inventor's intellectual property that he can use to evict trespassers (infringers). It also gives him something tangible that he can lease or sell. It defines the inventor's ownership and provides him with a formal legal document proving it.

Although there's some pride in getting a patent, obtaining one shouldn't be viewed as an end in itself. Most patents are entirely worthless. They either give protection to useless inventions or are of such limited scope that they offer no real protection at all.

Surprisingly, about two of every three applications filed in the USPTO result in the granting of a patent, so they're clearly not difficult to obtain. Often companies or individuals will boast about how many patents they have. That's not really a relevant productivity measure. The real question should be how many patents they hold that have resulted in profitable products. Anyone willing to pay the price can file and get nearly as many patents as he wants, but the patents might not be worth anything at all. In that case, the number of patents one holds could serve only to demonstrate how foolishly he has wasted his money.

Only a very small percentage of patented inventions ever make it to the marketplace. Of those, even fewer become profitable. But don't be put off. Most failures stem from starting with an unsound idea, from having too little capital, or from poor business management. As we go along we'll discuss ways to minimize these failure modes.

If you can establish and protect the ownership of a fundamentally good idea, you have a very good chance of cashing in. Couple that with frugality and hard work and you'll get there. The important thing is to keep anchored to practicality.

There follow descriptions of three patent types: *utility patents, design patents,* and *plant patents.* In addition, there are *provisional patent applications,* which are useful in many circumstances. As mentioned earlier, they provide an inexpensive way to get some minimal, temporary protection.

A very good summary guide to the patenting process is found at:
http://www.uspto.gov/patents/process/index.jsp.

Utility Patents

Utility patents are the most desirable and powerful patent type. They cover new and useful processes (or methods), machines, manufactured articles, and compositions including recipes, drugs or other chemicals, or any new and useful improvements to them. To qualify for a utility patent, in addition to falling into one of the above categories, the invention needs only to satisfy the following three conditions:[2]

The invention must be useful. That is, it must perform some function.

The invention must be novel. That means it cannot have been previously

- known or used by others in this country, or patented or described in a printed publication in any country, before the invention thereof by the patent applicant; or,
- patented or described in a printed publication in any country or in public use or on sale in this country more than one year prior to the effective filing date (EFD)[3] of the patent application.

If the inventor describes the invention in a printed publication or uses the invention publicly, or places it on sale, he must apply for a U.S. patent before one year has gone by, otherwise any right to a patent will be lost. Anything that has ever been publically known about an invention other than that disclosed by the inventor within the year preceding its EFD, is known as *prior art.*

The invention must be sufficiently different from what has been used or described before that it may be said to be non-obvious to a person having ordinary skill in its technology area.

In addition to the above listed three conditions, to obtain a patent the invention and its use must be clearly and completely described in the patent application. Note that there's no requirement the invention be commercially interesting or even a good idea. It could be absolute trash. It's not the USPTO's role

2 For more detailed information see: http://www.uspto.gov/patents/resources/general_info_concerning_patents.jsp#heading-4

3 A more precise definition of EFD can be found in Appendix I.

to decide whether or not the invention's any good. A cursory issued-patent review makes that fact abundantly obvious.

Figure 4 is a good example. It shows a drawing from U.S. Patent 4,605,000. The invention is a "greenhouse" helmet worn over the head and affixed to the wearer's body by shoulder straps. It contains live plants such as the cactus shown in *Fig.2* of the drawing. The pots holding the plants are adhered to small shelves within the dome-like helmet by little pop-in nibs. The wearer breathes oxygen given off by the plants inside. Inflow and outflow filters allow outside air to pass through the dome. *Fig.1* of the patent drawing illustrates the helmet being worn by a jogger. The invention meets all of the above criteria to be granted a utility patent: It performs a function; it is certainly novel; and, it is without a doubt non-obvious.

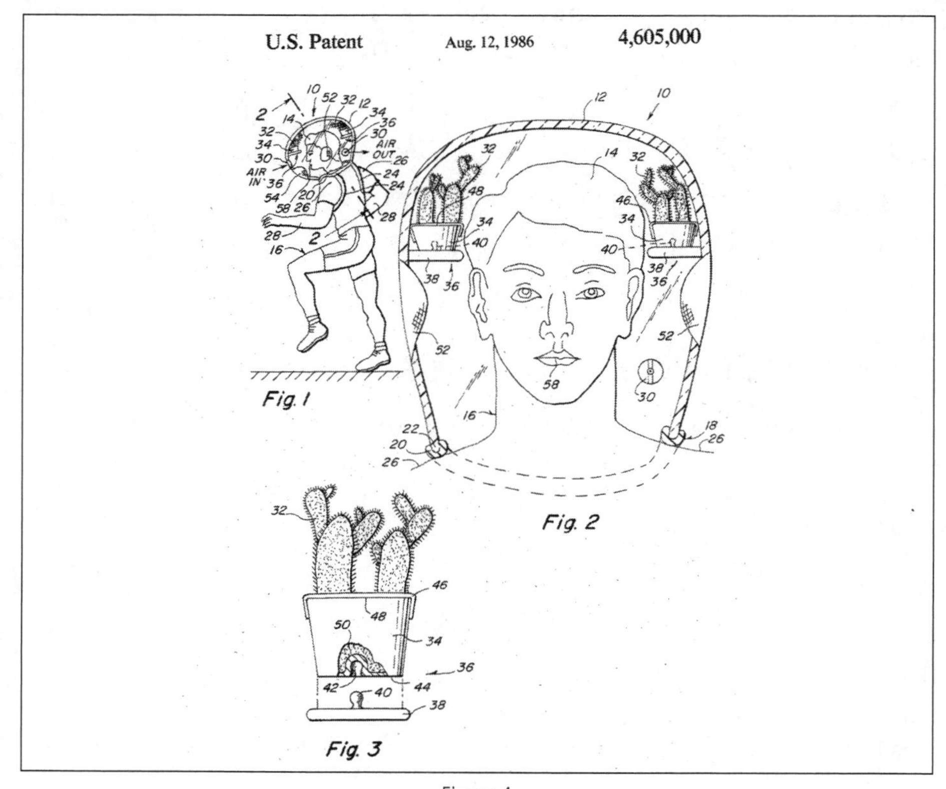

Figure 4

Can you imagine jogging with a dome full of potted cactus bouncing around your head. Better tie your shoes first. You won't be able to lean over once you put it on. Apparently the helmet was not a big success: The patent was allowed to expire due to non-payment of maintenance fees.

New utility patents have a 20-year[4] life, beginning on the effective filing date[5] on which they're applied for, assuming the required maintenance fees are paid in the middle of the third, seventh, and eleventh years after the patent's grant date. If the maintenance fees are not paid, the patent expires on the fourth, eighth, or twelfth grant-date anniversary, respectively.

Only the inventor(s) or his qualified legal representatives may apply for a patent. A patent applied for and granted to anybody else is invalid, and the applicant might be subject to criminal penalties. If the inventor cannot or won't file, a co-inventor can apply on their joint behalf. If there's no co-inventor, or if the co-inventor also cannot or won't file, anybody with a proprietary interest in the invention, such as a qualified employer, may apply on the non-signer's behalf. The inventors remains the named inventors of record in the patent. Nobody else can claim to be the inventor.

Please go to the Quiz on page 259.

You should now be able to respond correctly to quiz items 37 through 50.

4 Lifetimes of pre-1995 patents are the greater of seventeen years from the issue date or 20 years from the earliest filing date.

5 Please see Appendix I for a full definition of "effective filing date."

Design Patents

These cover new and non-obvious *ornamental, non-functional* features of manufactured articles. They protect only the article's appearance. Design-patent format is more or less the same as that for utility patents, but whereas a utility patent has a 20-year lifetime from the *application* date and permits many claims to uniqueness, a design patent's term is only fourteen years from *grant* date and permits only one claim. Since there's only one claim, design patents are fairly easy to get around, and they provide minimal protection.

Sometimes an invention can be covered by both a design patent and a utility patent and, of course, that's superior. As a dual-patent example, a better toaster could work exactly like the classical toasters of old, but have a unique appearance; maybe it's formed to look like a covered wagon. It would then be appropriate for a design patent. If it also toasted bread differently, it could be appropriate for a utility patent as well.

In another example, imagine a writing pen with a distinctive, easily recognizable shape that sets it apart from its competitors on the store shelf; perhaps it looks like a banana. But, suppose it also incorporates a novel point that distributes ink onto the paper in a way different from previously known pen points. Now the pen is eligible to be considered for both patent types: a design patent for the distinctive shape and a utility patent for the functional point.

There are no maintenance fees for design patents.

Plant Patents

These cover distinct, new asexually reproduced plant varieties other than a tuber-propagated plant or a plant found in an uncultivated state. The formats for plant patents and plant-patent applications are the same as those for utility patents except that the drawings may be rendered artistically and can be in color. Usually, color photographs are used.[6] The plant patent's lifetime is the same as a utility patent's: 20 years from the filing date. A plant patent is granted on the entire plant, and only one claim is permitted.

There are no maintenance fees for plant patents.

Provisional Patent Application

Provisional applications can be filed for utility and plant patents but not for design patents. They provide a fast, low-cost U.S. patent application alternative. They establish an early effective filing date for a later traditional, or "nonprovisional," patent application and permit the term "Patent Pending" to be used in connection with the invention. There is no such thing as a "provisional patent," as the "provisional" application itself never results in a patent. It's a provisional version of the nonprovisional patent application that might be filed later.

The provisional patent application's priority date is the date on which it's filed in the USPTO. The applicant then has up to 12 months to file a nonprovisional patent application in order for the claims in the later-filed nonprovisional application to benefit from the filing date of the provisional application. The provisional patent application itself can be filed with or without claims. The claims only benefit to the original filing date of the provisional to the extent that they are supported by its content. Provisional applications are not examined for merit, and they're abandoned by law 12 months from the filing date. The time between the provisional patent's filing and the subsequent

6 Colored figures are permitted in some circumstances for utility patents as well, but only when absolutely necessary, and that's not often.

(within 12 months) nonprovisional application's filing isn't counted toward the 20-year patent term.

Provisional patent applications offer some protection to the inventor, permitting him to discuss his invention with potential partners, vendors, manufacturers, etc. Care should be exercised, however, as the inventor's ownership scope isn't fully determined until the subsequent patent resulting from a nonprovisional application is issued. Remember, the patent application's proposed claims are frequently not those that will ultimately be granted.

The provisional patent application's details should be kept secret by the inventor. To the extent he discusses his invention with others in this phase, he should only do so under a non-disclosure confidentially agreement. The application's content remains his trade secret until after the nonprovisional application is published. The USPTO will publish the nonprovisional application 18 months after the filing date of the provisional application.[7] The provisional application is never published. Take note: the patent's 20-year lifetime starts with the *nonprovisional* application's filing, but the eighteen month secrecy interval preceding publication starts with the *provisional* application's filing.

The provisional application's principal advantages are that it lets the inventor get something formal on record and buys him a little time to advance his program before laying out the cash for a nonprovisional filing. In that time, he can do more toward proving his concept and perhaps attracting funding and customers. If he finds in the provisional stage that his invention isn't working out as he planned, he can abort with minimum losses.

If the inventor still isn't ready to file the nonprovisional application at the end of the 12-month lifetime of the provisional application, he can simply abandon

7 The 18-month publication can be avoided at the inventor's request, providing that he elects not to file any foreign patents on the invention.

the provisional application and file a second one on the same subject matter. In doing so, he will lose the priority date of the first provisional application. It's as if the first provisional application never existed. Keep in mind, though, that this can only be done if the invention has not been publically disclosed more than 12 months prior to filing the second provisional application.[8]

The filing of a provisional patent application does not constitute public disclosure of the invention.

Notices

Patent owners must ensure their patented articles are marked with the words "U.S. Patent" and the patent number(s). The penalty for failure to mark is that the owner(s) may not recover damages from an infringer unless the infringer was duly notified and continued to infringe after the notice. It's best to be very diligent about appropriately marking your products. It's so easy to do, why complicate life unnecessarily by failing to mark your products?

The terms "patent pending" and "patent applied for" indicate either a provisional or nonprovisional patent application on that article is on file in the USPTO. Those who use these terms falsely are subject to fines.

Please go to the Quiz on page 259.

You should now be able to respond correctly to quiz items 51 through 61.

8 A later section of the book entitled "Foreign Filing" describes a way of combining a provisional patent application with the filing of a Patent Cooperation Treaty (PCT) application to stretch out the costs of patent filing.

Introduction to Trademarks

A *trademark* is a word, name, symbol, or other identifier used by its owner to identify and distinguish his products from those of others. A good trademark can be a valuable asset to a business. Take McDonald's Golden Arches logo as an example. The well-known symbol is recognizable worldwide.

The first person to use a trademark in a given geographic market owns the trademark in that market. Common-law rights to a trademark are established simply by correctly using it. You don't have to register it; however, a Federal Trademark Registration provides several advantages, including the following:

- Notice to the public of the registrant's ownership claim of the mark;
- A legal indication of the registrant's exclusive right to use the mark ***nationwide*** on or in connection with the goods and/or services listed in the registration;
- The use of the U.S. registration as a basis for obtaining registration in foreign countries; and,
- The ability to file the U.S. registration with U.S. Customs and Border Protection to prevent importation of infringing foreign goods.

Once you establish rights to a trademark, you may use the "TM" designation to alert the public to your claim, regardless of whether or not you have filed an application with the USPTO. But if you do file a federal trademark registration, once it is registered you may then use the "®" symbol on all goods and services listed in the registration. Trademarks, unlike patents, can be renewed perpetually as long as they're being used commercially.

U.S. trademark applications are based on an intention to use the mark in commerce or actual use of the mark in commerce. An "Intent-to-Use" application can be filed as soon as you have decided on a name, slogan, or logo, and it reserves protection for the mark starting on the date you file the application, provided that you follow all of the requirements through to registration of the mark, including ultimately making use of the mark in commerce. Use of a mark in commerce requires use of the mark in the normal course of business; that is, selling your product or service to a customer in the normal course of trade. A sale to a friend just to establish "use" does not qualify and can invalidate your trademark. When in doubt, file an Intent- to-Use application.

Federal trademark registration is not straightforward. One problem is that many trademarks are very similar, and so being sure yours does not infringe someone else's is difficult. Additionally, since it's not required that trademarks be federally registered, others might have legal common-law rights to trademarks that would not show up in a search of the USPTO's database. The strength of a trademark is based on how "distinctive" the mark is. Descriptive names are not subject to trademark protection (e.g. "wood" for a table or "cold" for ice cream). Names that are not descriptive can be protected, if there are no other marks with the same or very similar content (e.g. Chili's restaurant, Dove Soap, or Delta airlines). The strongest trademarks are terms that are completely made up, like Buick or Kodak.

Unless you're willing to really do some heavy research on your own, it's probably better for you to use the services of an attorney to register your trademark. An attorney's thorough search of all trademark data bases, probably something you cannot readily do yourself, could save you from future complications in the use of your mark, and possibly even from costly legal problems.

Introduction to Copyrights

Copyright protection covers only the tangible expression of an idea, not the idea itself. The subject is therefore of less general interest to inventors than the protection of their ideas. Copyrights protect creative works such as writings, music, and works of art, and are obtained automatically when the work is reduced to tangible form. Copyrights do not have to be registered. But they can be, and there are some advantages to doing so, particularly if a downstream legal challenge is anticipated. Copyrights are registered in the U.S. Copyright Office, which is part of the Library of Congress. It's easy and relatively inexpensive to register your copyright. Under many common circumstances copyright protection lasts for the lifetime of the author plus 70 years, but there are exceptions to that. To learn more about copyrights, including how long the protection lasts, and how to register yours go to: **http://www.copyright.gov/.**

The website has an almost overwhelming amount of information, more than you will ever need to know to protect your work. But it also has some very good, simple tutorials for beginners. When you go to the website you will find a large number of clickable topics. You could start by clicking on "Taking the Mystery out of Copyright." That leads to a simple four-part presentation on copyright basics. After the presentation you will have a good idea of how the process works, and can go on to the site's more detailed information if you want to learn more.

Introduction to Trade Secrets

A *trade secret* is any proprietary method, procedure, drawing, or other material which describes the invention, its manufacture, or its assembly and is held in confidence by the originator.

Trade secrets provide a way to protect your intellectual property without filing any documents whatsoever with the USPTO. In most circumstances, trade secrets and patents are combined in such a way that the patents protect the basic invention, and the trade secrets protect the manufacturing know-how. However, there are some circumstances in which trade secrets alone are a protection form superior to patents and may altogether eliminate the desirability of obtaining a patent.

As an example, suppose a certain manufacturer has discovered a new way to economically produce seamless metal tubing of unequalled quality, and there's a profitable high-reliability market for it. No one else can figure out how to do it. The manufacturer could file for and probably receive a patent for the method.

Thereafter, for 20 years no one else could use it without infringing. But, in publishing the patent, he would have to reveal the method. Now a large competitor learning his method from the patent might just go ahead and infringe, resulting in expensive litigation lasting for years. The inventor might win something in the litigation, but his customer base could be ruined by the competitor and he could be financially destroyed.

In another scenario, having learned his method by way of the patent, a different manufacturer might see ways to modify or even improve it so as to avoid his claims entirely and still make a competitive product.

Also, in filing the patent, the inventor has limited his protection to 20 years. After that, anyone could use his method.

So, in a case like that of the tubing manufacturer, the inventor should probably choose simply to keep his method secret. As long as he doesn't reveal his method in any way, it can be used by no one else. Even if someone learns his method in a fraudulent way, it cannot be legally used. The manufacturer does have the obligation to strictly maintain his method's secrecy, however. And, if he reveals it even once to anyone without properly notifying them of its proprietary nature, it moves into the public domain.

There are many well-known examples of trade secret protection, the most famous of which is probably the recipe for Coca-Cola. Had Coco-Cola patented its recipe, after twenty years the patent would have expired, and the recipe could have been copied by anyone. But as long as the company maintains the recipe's secrecy, it's theirs perpetually.

Trade-secret maintenance requires constant vigilance. All drawings, procedures, and any other documents referencing the proprietary information must be tightly controlled and appropriately marked. Areas where proprietary processes take place or proprietary documents are kept must be off-limits to casual visitors. Employees should sign confidentiality agreements that bind them to never disclose your trade secrets, even after they leave your employment. Clients, vendors, and other business associates who might, with or without consent, acquire proprietary information should be required to sign similar non-disclosure agreements.

Trade-secret holders cannot be too careful. Only one slip is needed to lose it all. If anyone not bound by a confidentiality agreement learns the proprietary information in a legal way, it's no longer a secret.

An intellectual property attorney should be consulted about appropriate security precautions such as markings and notices.

The foregoing has been an introductory overview of patents, trademarks, copyrights and trade secrets. What follows is a more in depth opportunity to study patents.

Please go to the Quiz on page 259.

You should now be able to respond correctly to quiz items 62 through 71.

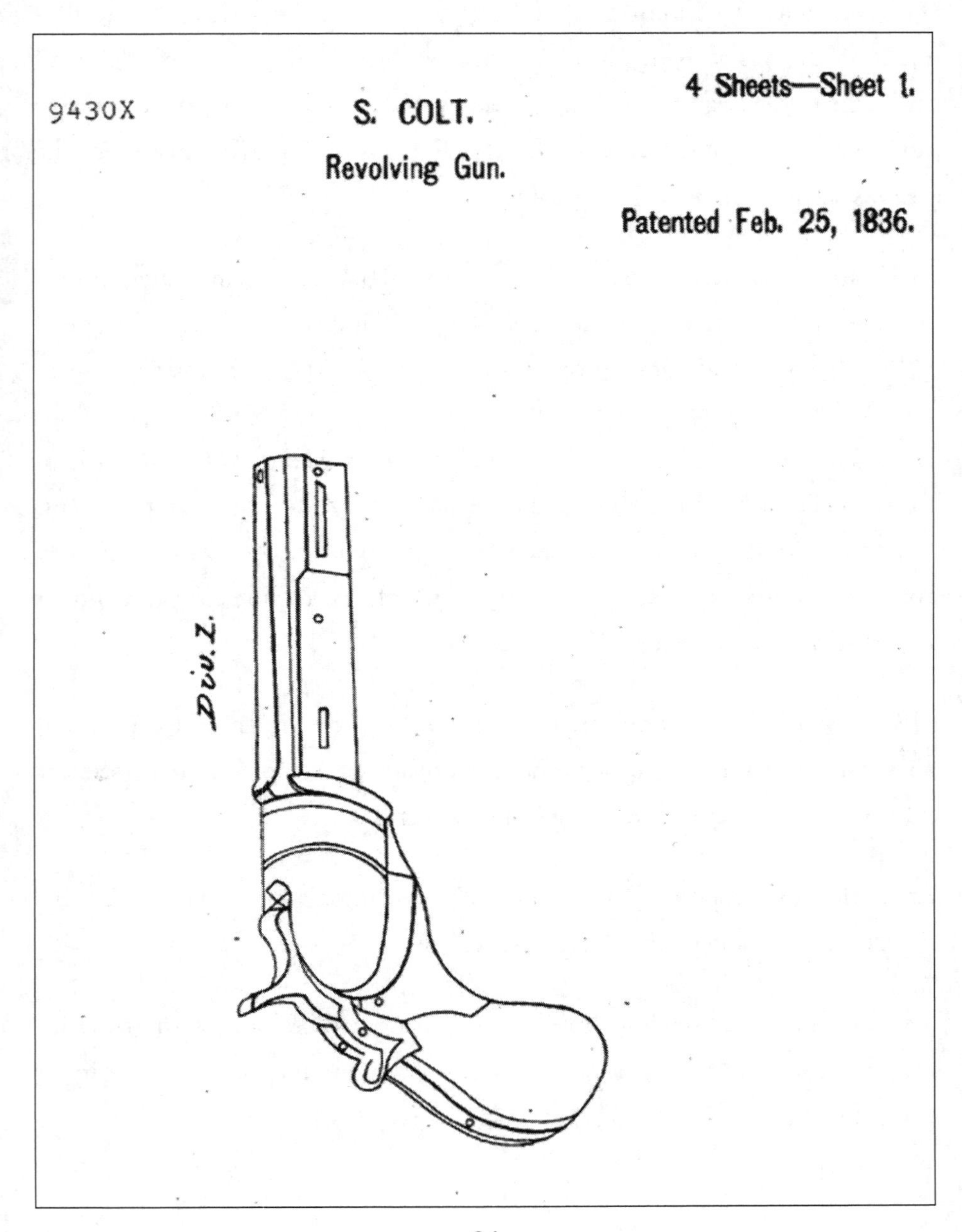

PATENTS: IN-DEPTH STUDY

A wealth of comprehensive, in-depth information can be found on the USPTO's official website:[1] **www.uspto.gov.**

The first time you go to the site, allow yourself a few hours to navigate around and become familiar with it. The U.S. Government's intention to support and nurture innovation will be abundantly apparent to you as you visit the site's many useful sections. It is user friendly; still, there's such an enormous amount of material there that it takes a lot of time to explore. But it's time well spent. It's extremely interesting, and it's free. What more could you want? Don't just read what is written here about the various website learning opportunities; actually visit the suggested sites, carefully watch the tutorials and take the quizzes. Again, learning to be an inventor takes a lot of work; so dig in and seriously study what follows.

1 Not to be confused with www.uspto.com, which is a non-government commercial site

I will lead you to the most relevant parts of the USPTO site to get you started. We begin by looking at the sections dealing with patents. Keep in mind the website changes from time to time, and the routine suggested here will probably have been modified by the time you go there. But it is current as of the time of this writing, and should, in any case, give you a good foundation for using the site.

Here are the steps to follow:

- **Go to www.uspto.gov.**

  There is a clickable icon titled "INVENTORS." Click it.

  The icon leads you to a web page called "*Inventor's Resources.*" It is devoted to independent inventors and offers educational material covering all patent and trademark aspects. The site attempts to answer all the beginning inventor's most common questions. There is a list of topics in a blue background menu on the upper left hand side. You will come back to this menu repeatedly in the following discussions.

- **Click on the "Patents for Inventors" menu item.** It leads you to a comprehensive discussion of U.S. patents. It contains brief discussions of just about everything an independent inventor needs to know about patent basics, including how to apply for one.

  Back on the Inventor's Resources page, next click on the "*Education and Information*" menu item. It opens a page full of useful information. Start by clicking on the "*From Concept to Protection*" title. That opens a worthwhile lecture on how to begin protecting your intellectual property from the very start of your work. When you have finished the lecture, go back to the Inventor's Resources page.

There is one piece of advice in that lecture which should be followed with great care, if at all. It advises you to get someone to sign the pages of your invention logbook as a witness. If you do that, keep in mind that the witness might suggest some patentable change to your invention, and you could have an unwanted co-inventor. Or, the witness could fail to keep your developments secret.

There is another caution about having a witness. Recent changes in patent law grant ownership to the first inventor who *files* a patent application on the invention.[2] Previously it was granted to the one who first *invented* the invention. If your witness were to file on your invention before you do, you could be faced with litigation to establish ownership of your own invention. It's true that the witness could not legitimately file on your invention, because only the true inventor(s) may do so; but if he fraudulently did so, you would be faced with having to disprove his claim. So if you get someone to witness your logbook, be sure to get him to sign a non-disclosure agreement beforehand stating that he has not participated in the invention. That at least gives you some protection. Clearly, when you have someone witness your work, you are putting a whole lot of faith in that person's integrity. Using your mom as a witness seems more appealing all the time.

- **Next click on the "Pro se-Pro bono" menu item.** This one is still in its formative stage, but it already contains a lot of interesting information. Start by clicking on the "*Pro Bono*" title, where you'll find a training video explaining how, under certain circumstances, inventors can get free legal assistance in filing U.S. patents.

2 For more details see the later section entitled: More Decisions!. The Pathway beyond Gate 8

The "*Pro se*" portion is under construction while being updated. When finished, it will contain training modules on the patent process and on patent searching techniques, sample provisional applications and direct links to forms and fees needed for filing a patent application.

- **Next, go back once again to the Inventor's Resources page and click on the "Inventor's Assistance Center" menu item.** It leads you to a list of Inventor Assistance Center phone numbers you can call for one-on-one assistance with your patent questions. Believe it or not, real people actually can be easily reached by phone there. It also gives a web address for the Patent Ombudsman Program, which is there to resolve any issues arising in the patent prosecution process.
- Local inventor's resources can be found by going to the Inventor's Resources page, and clicking on the "*State Resources*" title. Here you can find additional links to local USPTO-registered patent attorneys and agents, U.S. Patent and Trademark Resource Center (PTRC) libraries, and inventor's clubs. The list of registered attorneys and agents will help you find professional intellectual property assistance in your area. As you will have learned from the lectures recommended earlier, for an attorney or agent to represent you before the USPTO he must be registered.

The PTRC libraries house comprehensive hardcopy U.S. patent collections and related materials. The location of the PTRC library nearest to you can be found online at: **http://www.uspto.gov/products/library.**

When you're on that web page, click on "Locate Libraries" and follow the trail to get the library locations.[3] Before there was so much well organized material available on the Internet, it was necessary to go to such a resource center to research patents. Nowadays, however, that is rarely, if ever, required.

Inventors Organizations: There are many independent inventors clubs across the United States. A list of them can be found at: **http://www.uspto.gov/inventors/independent/eye/201112/orgs.jsp**

If you are not already associated with your local inventor's group you should look into it. The inventor's clubs I have been involved with have monthly meetings in which the members share information, help each other with relevant problems, host lecturers, and enjoy the camaraderie of their shared passion for inventing. They offer a good avenue for networking to local inventor assistance.

Still another useful patent website:

The intention in this section of the book is not to help you become an IP expert, but to lead you to sources that will give you a good overall understanding of the subject. So, just when you might have thought you were finished studying patents, here's another excellent U.S. Government resource for clear patent information that you should look into. The following website is tailored for independent inventors and small businesses: **http://www.stopfakes.gov/business-tools.**

The site's material is presented in a slightly different format than that of the USPTO site; it's well worth studying. When you first go to the site, click on

3 You might try going directly to: http://www.uspto.gov/products/library/ptdl/locations/index.jsp for the library locations. This portion of the USPTO website has been under change recently; right now it works.

"Business Tools." There you will see a list of topics, two of which are particularly useful. To get to them:

- Under the "*IPR Basics*" heading click on the "*Learn about IPR*" menu item. That will open a new page titled "*Learn about Intellectual Property*" which gives a clear, uncomplicated discussion of patents, copyrights, trademarks, and trade secrets. It has answers to many frequently asked questions.
- On the same *IPR Basics* webpage there's a clickable icon titled *"IPR Training Module."* Click on that. Then click on the language you wish to use. That opens an online intellectual property training course that takes about an hour and a half to complete. It can be completed online, or is available in a printable PDF version. If you are a member of an inventors association, I suggest you download the PDF version and go through it as a group.

Once you have studied your way through the information on the two U.S. Government websites just discussed you should have a good basic working knowledge of patents. In later sections certain related topics, such as patent searching, patent claims analyses, patent filing, and others will be discussed in some detail. As a prelude to that, we will first look at a utility patent chosen for its simplicity.

A Sample Utility Patent:

The following pages illustrate a brief patent for an uncomplicated bottle opener. As we describe the USPTO's required patent and patent-application format, please refer to this example. Later, you can use it and the accompanying comments as guidelines for preparing your own nonprovisional or provisional patent application's first draft.

United States Patent [19]

Halpin

[11] **Patent Number: 5,056,383**

[45] **Date of Patent: Oct. 15, 1991**

[54] **BOTTLE OPENER**

[76] Inventor: **Harold W. Halpin,** 2902 N. 44th La., Phoenix, Ariz. 85031

[21] Appl. No.: **670,563**

[22] Filed: **Mar. 18, 1991**

[51] **Int. Cl.**[5] **B67B 7/18**
[52] **U.S. Cl.** **81/3.43;** 81/64
[58] **Field of Search** 81/3.4, 3.43, 64, 3.07; D8/33, 40, 43

[56] **References Cited**

U.S. PATENT DOCUMENTS

D. 268,164	3/1983	Sandberg	81/3.43
2,132,207	10/1938	Donovan	81/3.43
2,317,037	4/1943	Donovan	81/3.43
3,084,573	4/1963	Lipski	81/3.43
4,150,592	4/1979	Mott	81/64
4,660,445	4/1987	Windom	81/3.43
4,889,018	12/1989	Shaffer	81/3.43

FOREIGN PATENT DOCUMENTS

1782406	8/1971	Fed. Rep. of Germany	81/3.43
647271	7/1928	France	81/64

Primary Examiner—Roscoe V. Parker
Attorney, Agent, or Firm—Richard G. Harrer; Charles E. Cates

[57] **ABSTRACT**

A twist-cap bottle opener having a handle provided with a bottle cap gripping member attached to one end of the handle. The bottle cap gripping member is generally circular shaped spring-like material, one end of which is attached to an end of the handle. The opposite end of the circular shaped spring material remains unattached and thus the diameter of the cap gripping member can be easily varied to accommodate different sizes of bottle caps or jar lids. The inner surface of the cap gripping member, that is the surface that is in contact with the side of the bottle cap, is provided with projections which act to grip the side of the cap. The unattached or free end of the cap gripping member is bent inwardly to further increase the effectiveness of the opening device.

3 Claims, 1 Drawing Sheet

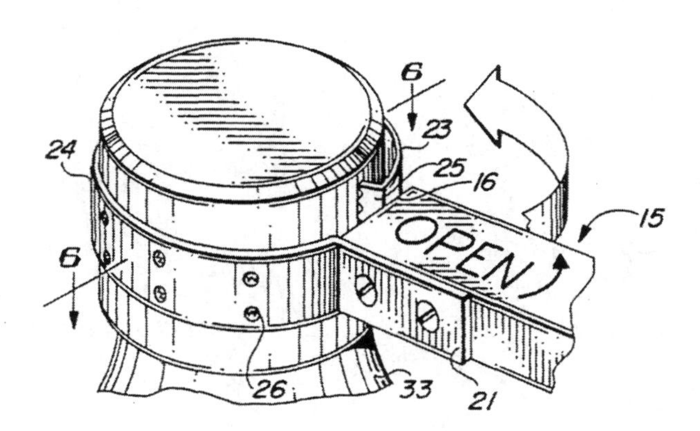

U.S. Patent **Oct. 15, 1991** **5,056,383**

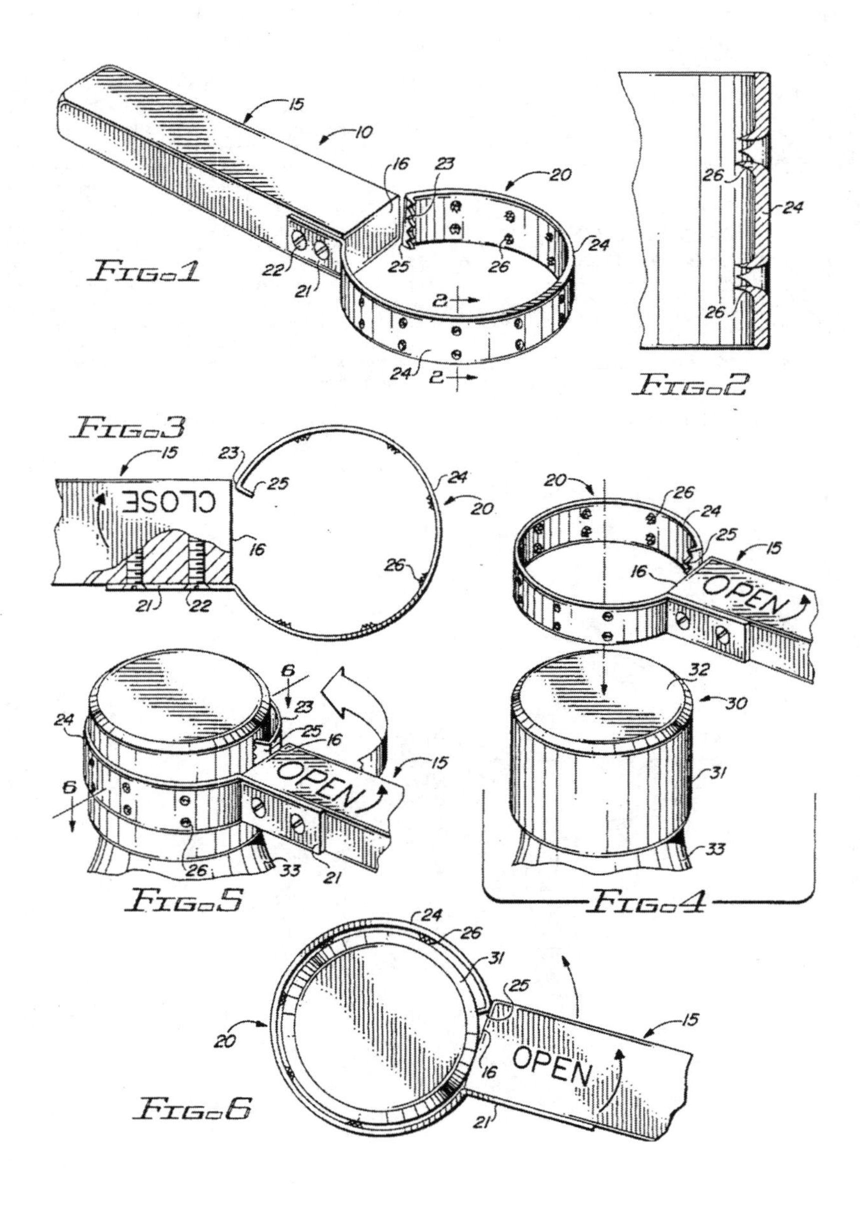

5,056,383

1

BOTTLE OPENER

FIELD OF THE INVENTION

The present invention relates generally to a device for removing bottle caps and more specifically to a bottle opener particularly adapted to be used in removing twist-on caps from bottles.

BACKGROUND OF THE INVENTION

In rather recent years, the use of glass and plastic bottles for beverages such as soft drinks has increased tremendously. Virtually all such bottles are sealed by means of a relatively small diameter screw-on or twist-on cap customarily made of metal or plastic. Such twist-on caps enable the user to remove some of the beverage from the bottle and then reseal the bottle to preserve the freshness of the beverage. Although such beverage packaging allows the user to reseal the bottle if desired, and is an efficient means of storing the beverage, such caps can be very difficult to remove by people who lack the necessary hand strength needed to remove the cap. Moreover, if the bottle is to be resealed, some may not have the necessary hand strength to replace the cap to prevent loss of carbonation. In U.S. Pat. No. 4,660,445 there is disclosed a device for removing twist caps on bottles, the device having an enlarged generally dome-shaped housing, with an adjustable clamp secured within the housing for circling various sizes of bottle caps. A remote adjustment handle extends through the housing for adjusting the clamp to fit various cap sizes. Further examples of strap-like wrenches for encircling and removing jar caps are shown in U.S. Pat. Nos. 3,084,573; 2,317,037; 2,132,207 and U.S. Pat. No. 268,164. Although the foregoing opening devices are probably all effective in removing caps or lids from jars or bottles, for the most part they are quite complicated in structure and likely to be rather costly to manufacture.

SUMMARY OF THE INVENTION

This invention provides a twist-cap bottle opener having a handle provided with a bottle cap gripping means attached to one end of the handle. The bottle cap gripping means is a generally circular shaped spring-like material, one end of which is attached to an end of the handle. The opposite end of the circular shaped spring material remains unattached and thus the diameter of the cap gripping means can be easily varied to accommodate different sizes of bottle caps or jar lids. The inner surface of the cap gripping means, that is the surface that is in contact with the side of the bottle cap, is provided with projections which act to grip the side of the cap. The unattached or free end of the cap gripping means is bent inwardly to further increase the effectiveness of the opening device.

BRIEF DESCRIPTION OF THE DRAWINGS

While the specification concludes with claims particularly pointing out and distinctly claiming the subject matter which is regarded as the invention, it is believed that the invention, objects, features and advantages thereof will be better understood in the following description taken in connection with the accompanying drawings in which like parts are given like identification numerals and wherein:

FIG. **1** is a perspective view of the bottle opener;

2

FIG. **2** is a sectional view taken on the line **2—2** of FIG. **1**;

FIG. **3** is a partial bottom view of the bottle opener;

FIG. **4** is a partial perspective view of the bottle opener about to be placed on a bottle;

FIG. **5** is a partial perspective view of the bottle opener positioned on the cap of the bottle and in position to remove the cap; and

FIG. **6** is a sectional view taken on the line **6—6** of FIG. **5**.

BRIEF DESCRIPTION OF THE PREFERRED EMBODIMENT

Referring now to FIG. **1**, the bottle opener which is shown generally at **10** includes a handle **15** provided with a cap gripping means shown generally at **20**. Cap gripping means **20** includes a generally circular shaped band **24** formed of a relatively thin, spring-like material, with end **21** of band **24** securely attached to one end of handle **15** by means of screws **22**. As shown, circular band **24** constitutes an arc of about 360 degrees and in any event it should form an arc of at least about 180 degrees. Band **24** is preferably made of spring steel having a thickness of about 0.016 inches. As show best in FIG. **2**, the side of band **24** is pierced to provide the inner surface of the band, that is the surface which contacts the cap, with a series of spaced apart and inwardly facing projections **26** which act as gripping teeth. The free end **23** of circular band **24** is bent inwardly at an angle of about 90 degrees and the very end of the band is then notched to provide additional gripping teeth **25**.

In use and to aid in the removal of a twist cap from a bottle, as shown in FIG. **4**, the cap gripping means **20** is centered over a bottle **33** having cap **30**. It will be seen that in this embodiment the word "OPEN" and an arrow is imprinted on one side of handle **15**. With this side of the handle in an "up" position, thereafter, as shown in FIG. **5**, the opener is lowered onto cap **30** so that band **24** of cap ripping means **20** surrounds the periphery **31** of the cap and is in contact therewith. To continue the procedure of removing the cap and as shown in FIG. **5**, the handle is grasped and turned in a counter-clockwise direction as indicated by the arrow on handle **15**. The shoulder **16** is in contact with the periphery of the cap. For removal of the commonly used twist caps on beverage bottles, the diameter of the cap gripping means **20** ranges from about 1 to 1½ inches. To remove the cap, the band **24** and its attendant projections **26** and teeth **25** should make good contact with the periphery of the cap and the shoulder **16** of handle **15** contacts the periphery of the cap. If the diameter of the cap gripping means is somewhat larger than the actual diameter of the cap, appropriate contact of the band with the cap may be made by using the fingers to reduce the diameter of the cap gripping means **20**. Since the band **24** is made of a spring-like material which has a considerable degree of flexibility, it is relatively easy for the user to reduce the diameter of the gripping means **20** by moving the free end of the band **24** in a clockwise direction.

After good contact has been made between the band and the periphery of the cap, movement of the handle in a counterclockwise direction will result in removal of the cap with considerably less hand strength required.

As will be seen in FIG. **3**, the bottle opening device of this invention can also be used to reseal a cap on a bottle and this is done by merely turning the device 180 de-

5,056,383

3

grees so that the word "CLOSE" and an arrow which is printed on the reverse side of the handle is readable. In this position, the handle is moved on a clockwise direction ensuring that a good seal is obtained by the cap on the bottle.

While this invention has been described in detail to particular reference to a preferred embodiment, it will be understood that variations and modifications can be effective within the scope of the invention as described hereinbefore and is defined in the following claims.

What is claimed is:

1. A tool for attaching and removing twist caps from containers comprising a handle, cap gripping means attached to one end of said handle, said cap gripping means further comprising a generally circular shaped band formed of a relatively thin, spring-like material, one end of which is attached to said handle with the opposite end thereof being unattached, the surface of said band which is in contact with said cap being provided with gripping means, with the unattached end of said band being bent inwardly to provide additional gripping means.

4

2. The tool of claim 1 wherein said band is made of spring steel and constitutes an arc of at least about 180° and wherein said unattached end of said band is bent inwardly at an angle of about 90°.

3. The tool of claim 2 wherein said band constitutes an arc of about 360° and wherein said gripping means includes a series of spaced apart inwardly facing projections.

* * * * *

Note: every section of the document's first page has a number appearing beside it in [square brackets][4]. The numbered sections are:

[19] United States Patent: This tells what the document is, followed by the inventor's last name.

[11] Patent Number: This is the number assigned to the granted patent. Patent numbers are assigned in increasing order. Patents with smaller numbers were granted before those with larger numbers. Patents issued at the time of this writing have numbers greater than 8,000,000.

[45] Date of Patent: This is the date on which the patent was granted.

[54] The Title of the Invention: It is called whatever the inventor wishes, as long as the title reasonably describes the invention. If it's a lawn mower, for instance, it couldn't be entitled "street lamp," but it could be called a "grass-cutting machine." The permitted latitude in naming inventions adds confusion to patent searching.

4 Not to be confused with the brackets used to designate points on the Map.

In our example, "Bottle Opener" closely describes the invention.

[76] Inventor: The inventor's name and address.

[*] Notices: There are no notices given in our exemplary patent, but many patents will have one or more sections inserted between the inventor's name and the patent application number [21]. They list information such as assignments or other details.

[21] Appl. No.: Each application is given an identifying number when it arrives at the USPTO.

[22] Filed: This is the date on which the application was filed in the USPTO.

[51] Int. Cl.[5]: United States patents are required to indicate the relevant, corresponding International Patent Classification symbols in accordance with an international agreement. These numbers are assigned by the USPTO and are not normally something the inventor needs to be concerned about.

[52] U.S. Cl.: This section lists the USPTO's domestic technology classes/subclasses into which the invention falls.

[58] Fields of Search: The U.S. patent classes and subclasses searched by the examiner in prosecuting the patent application are listed here.

[56] References Cited: Both the U.S. and foreign references cited by the examiner in preparing his findings are listed here. Generally speaking it is these references that have been cited as relevant close, existing art. They have limited the scope of the patent's allowed claims. ("Art" is the term generally given to the body of knowledge associated with a certain subject. "Old," "prior," or "existing" art is that which is already known before the present invention.)

The first reference cited in our bottle-opener example has a patent number preceded by the letter "D," indicating it's a design patent. The USPTO examiner's name and the name of the attorney, agent, or firm representing the inventor are given at the end of the reference section.

[57] Abstract of the Disclosure: This briefly describes the invention, including what there is about it that's new within the art to which the invention pertains.

The bottle-opener abstract accurately describes the innovation. It further lists advantages such as: "...easily varied to accommodate different sizes of bottle caps or jar lids" and "...free end of the cap gripping member is bent inwardly to further increase effectiveness."

The patent's remaining sections are not numbered. Each page is printed in two sequentially numbered columns. Each line within the columns is numbered from the page's top, and every fifth line is given an identifying number. When you prepare your patent application, you do not have to follow the page layout format. That is done by the patent office.

Background of the Invention: This section puts the invention in context with other means for accomplishing the same or similar functions. It states briefly how the invention works and why it's an improvement over existing art. In this section the invention's novelty, utility, and objectives are introduced.

Our bottle-opener patent's Background section clearly describes the bottle opener in context with others that are known and that perform a similar function. It states that the present invention is an improvement due to the facts that those already existing are relatively complicated and expensive. Our example patent has a "Field of the Invention" opening paragraph that could equally well have been included in the Background section.

Summary of the Invention: Here the invention's main points are summarized to give a brief but general understanding of what it's about. It prepares the reader to understand more easily the detailed specification that follows. It sets out the invention's main objectives.

The bottle-opener patent's summary section neatly describes the new opener and clearly states the invention's main objectives.

Brief Description of the Several Views of the Drawings, if Any: This section describes the drawings themselves, not the invention. It should give a title to each drawing and note, for instance, that a particular drawing is a cross-sectional view, a side view, or whatever it happens to be. Patent drawings must conform to certain specified rules available on the USPTO's website. The inventor, if capable, can make his own drawings according to the rules. Even if the inventor is not skilled at drafting, it's worthwhile for him to take a stab at the drawings. It gives the attorney or agent some guidance as to the most advantageous and clarifying views. Patent applications can be filed with informal, provisional drawings that must be upgraded before the patent issues.

In our example, the new bottle opener is described by several views that collectively illustrate all of its important features. Note that each feature subsequently mentioned in the patent's text is uniquely numbered on the drawings and that the numbers are referred to each time the feature is mentioned in the specification.

Specification of the Invention: Although usually referred to as the "Specification," or "Detailed Description," in our patent example this section is entitled "Brief Description of the Preferred Embodiment." The specification describes the invention and how it's made and used. It distinguishes the invention from other inventions and from old art. It completely discloses the invention's best embodiment as contemplated by the inventor and explains how it works.

In the case of an improvement, the specification points out the invention's part or parts to which the improvement relates and explains how the improvement works.

The specification describes the invention in the context of the patent's drawings. Both the written description and the associated drawings should omit features that are unnecessary to the invention's complete explanation. Additional, non-essential details make everything more complicated and expensive to present and harder to understand. All necessary features, however, should be presented and clearly defined.

Claims: The specification concludes with a claims section. It is the most important element of the application. More than one claim is allowed provided they differ from each other. Claims may be presented in independent form (e.g., the claim stands by itself) or in dependent form. Dependent claims refer back to and further limit another claim or claims in the same application.

In our bottle-opener example, the inventor has been granted three claims, the first of which is independent, the second dependent on the first, and the third dependent on the second.

The claims section is the patent's nucleus, as it defines exactly what's patented. The claims should clearly delineate all the invention's unique aspects. It's the claims that define scope of the patent's protection coverage.

The claims are interpreted in the context of the accompanying drawings and description. It's pretty much true, though, that if it's not in the claims, the inventor doesn't own it.

A Bogus Patent Example: Claims Discussion:

It's particularly important that you understand how patent claims are interpreted and written. You are encouraged to use a patent attorney or agent to write the application's claims, but you must carefully read and understand them. It's you, and not he, who should be most aware of the new technology's advantages and potential applications. Many patents, even for good, innovative technology, are virtually useless because the claims do little or nothing to protect the design's most important features. One wrong or neglected word in the claims can drastically reduce the patent's scope. The following somewhat ridiculous example illustrates the importance of precise wording.

Imagine you've come up with a new improvement for traffic stop signs. You couldn't patent a basic stop sign; that's clearly already in the public domain. But, suppose you think that to make the signs more evident you could add strings of dangling reflectors that would turn in the slightest breeze, twinkling reflected light. Then you could go on to describe how the reflectors would be suspended from the sign, their optimum size, shape, color, finish, etc. You could probably patent that concept and get at least one claim approved for each particular feature.

You might initially ask for broad claims (your wish list) such as the following:

I hereby claim:

1. A sign with at least one environmentally actuated device to enhance awareness of it.
2. A sign as in claim 1 in which the enhanced awareness is visual.
3. A sign as in claim 1 in which the environmental actuating agent is the wind.

4. A sign as in claim 2 in which the at least one actuated device is a movable reflector.
5. A sign as in claim 4 in which the at least one movable reflector is spoon shaped.
6. A sign as in claim 4 in which the at least one movable reflector is red.
7. A sign as in claim 4 in which the least one movable reflector is suspended from a flexible element.
8. A sign as in claim 7 in which the flexible element is a string.
9. A traffic stop sign with at least one environmentally actuated device to enhance awareness of it.

You'll note that Claim 1 is an independent claim standing on its own, but the claims 2 through 8 are dependent ones that refer back to other preceding claims. These dependent claims could go on and on. Claim 9 is an independent claim limiting the sign for use as a traffic stop sign.

If you were granted a patent with the above claims, you would then have a patent with multiple claims for a fancy sign. It would be useless, of course, but because this is only an example it doesn't matter.

Someone else could see your patent and think: "He missed the best part." Then he could go on to file his patent application for a sign with dangling strings of bells to add an audible signal. If your Claim 1 were approved, he couldn't get a claim on the audible angle. Your claim is broad enough to encompass a sign with any sort of ancillary, environmentally actuated device to enhance awareness.

But in truth, no competent USPTO examiner would allow your Claims 1 through 4 as written, because advertisement signs with sparkly sequins are common. So, you would have been constrained to rewrite your first claim to encompass independent Claim 9 in order to narrow it to stop signs, or something of that nature. Your competitor with the audible devices would still be foiled by your new Claim 1 even if you narrowed it to stop signs, because new Claim 1 doesn't specify visual awareness. Thus, it would still cover environmentally actuated bells, whistles, or whatever on stop signs.

Let's carry on the above example just one more step to see what effect a poorly worded first claim could have. Your initial concept was to use the environment to actuate something in order to enhance awareness of a stop sign. That's your fundamental idea. Your described preferred embodiment does it visually with reflectors. Just suppose for the moment that there would be a real market for a stop sign with an environmentally actuated audible signal…bells, for instance. Further suppose that your representative had worded your first claim too narrowly, such as the following:

I hereby claim:

1. A stop sign with environmentally actuated devices to enhance visual awareness of it.

Now you would have been cheated. True, the claim describes what you've chosen as your idea's preferred embodiment, but your first claim is unfortunately limited to visual awareness, and the competitor who takes your idea and replaces your reflectors with bells or whistles could get a patent on a device that is not part of what you own. Your patent application's author would have given away an important aspect of your idea. A major error involving one word: "visual," would have been committed.

Let's go back for a moment to our earlier Halpin 5,056,383 bottle-opener example. You'll see that his independent Claim 1 is very detailed. As noted in our earlier discussion, Halpin couldn't have been granted a broad first claim such as "A tool for attaching and removing twist caps from containers."

Other devices that accomplish the same function were already known, so every word following "containers" in his issued first claim has been added to distinguish his device from the others and permit the patent's granting. But, all those added words also narrow the protection afforded by the patent and reduce its value. They must be chosen very carefully. As a rule-of-thumb, the more words a patent's claim contains, the less it protects.

You can see from the foregoing examples that the application's claims portion is the one requiring the most skill to write. If not done carefully, it can leave the technology's broadest applications up for grabs, or it can fail completely to protect your invention.

While it's true that a professional application could be one of your major early expenses, if you do it yourself you could throw it all away with one poorly worded sentence. That's a risk not worth taking. Get professional help.[5]

Your first patenting step will most likely be to file a provisional patent application. They are less formal than nonprovisional ones, and more affordable: somewhere in the range of $500 when filed by an attorney or agent, and about $75 if you do it yourself. They need not necessarily even include claims. They should simply describe in as much detail as practical the invention's principal innovative features and how they function. Informal drawings, at least, should be included for clarity. As mentioned earlier, your provisional patent application is not examined by the USPTO until after a nonprovisional application on the same invention is filed. Then, the provisional application is scrutinized to determine that its content is supportive of the nonprovisional application, hence establishing the nonprovisional's earlier priority filing date.

5 There are self-help books that show you how do many of the procedural steps yourself, but I still do not recommend you do that. I advise the use of legal professionals in many aspects of your project. That's costly, and I don't like it any more than you do; but it's the surest way to protect your interests. The do-it-yourself alternatives are likely to leave you vulnerable.

Your next patenting step following filing the provisional application will probably be to file a nonprovisional application. Right now, a typical nonprovisional utility patent application filed by an attorney or agent costs around $10,000 to $15,000 depending on how complicated the invention is and how much you do yourself. In addition to fees, accompanying forms, and some ancillary information, a nonprovisional patent application has the same basic descriptive information that the corresponding patent, when issued, will have.

The $10,000 figure is punishing, and unfortunately it's a major impediment to independent inventors who typically cannot afford it. Nearly all of that money goes to the law firm who prosecutes the application. The only way around that is to file on your own behalf, but that has all the drawbacks just discussed. It's a bad situation all around because those high legal fees keep a lot of good technology from ever making it into public use; consequently, they are very detrimental to our nation's economy.

No matter how you accomplish it, once your non-provisional application is filed it goes into the examination queue. Upon initial examination, exception will probably be taken to some of the application's claims, or at least to some of their wording. Often, claims are rejected due to lack of novelty or because the claimed innovation is judged obvious in light of previously known technology. Additionally the examiner might note errors or drawing inadequacies that need to be corrected before the patent can be issued.

He'll issue a response called a "first office action" to your representative. It will state what in your application is acceptable and what's not. Working with you, your attorney or agent will try to amend it to satisfy the examiner's complaints while still meeting your needs. Remember, the application is your wish list. The issued patent is what you get. You usually don't get all you wish for. The back-and forth communication with the USPTO will continue until your application is approved or finally rejected. The dialogue could easily add a year or more to the period in which your patent is pending. Presently, it takes on average 22 months from the time a nonprovisional patent is applied for until it is granted.

In the end, you might feel a little beaten up by the process and probably won't have received the broad claim protection you initially hoped for. But, as mentioned earlier, chances are two out of three that you'll get a patent on some aspects of your invented technology. Once the patent is issued, you'll have to decide if its protection is adequate to justify further investment in the product.

In summary, you'll have made a decision at some point whether to file a provisional patent application or a nonprovisional one. As mentioned earlier, starting with a provisional application is advisable, particularly if you need a bit more time to research your invention's feasibility or marketability, or if you need funding or other support to advance the project. In any case, with either filing type you will have acquired some measure of protection for your invention and can begin to talk to others about it.

Remember, your invention remains your trade secret until the application is published or until you've otherwise publicly disclosed it. Until then, you should carefully disclose only those details necessary to accomplish your purposes. Do so under cover of a non-disclosure agreement. Even after publication, be sure to keep the invention's manufacturing know-how your trade secret.

As soon as your application is published, you have rights that provide you the opportunity to obtain a reasonable royalty from an infringer under two conditions:

- You must formally put the infringer on notice.
- A patent with a claim substantially identical to that infringed must be granted from your application.

In case you or any other invention stakeholder becomes aware of infringement, you should consult an attorney to assist in obtaining a remedy. Infringement is determined primarily by the issued patent's claims. If what the defendant is making does not fall within the language of your patent's claims, there's no infringement.

We'll have much more to say about patent filing and other patent matters in later sections. For the moment, you have acquired enough information to make the decisions that await you beyond the gate at point [8].

Please go to the Quiz on page 259.

You should now be able to respond correctly to quiz items 72 through 80.

2 Sheets—Sheet 1.

G. B. SELDEN.

ROAD ENGINE.

No. 549,160. Patented Nov. 5, 1895.

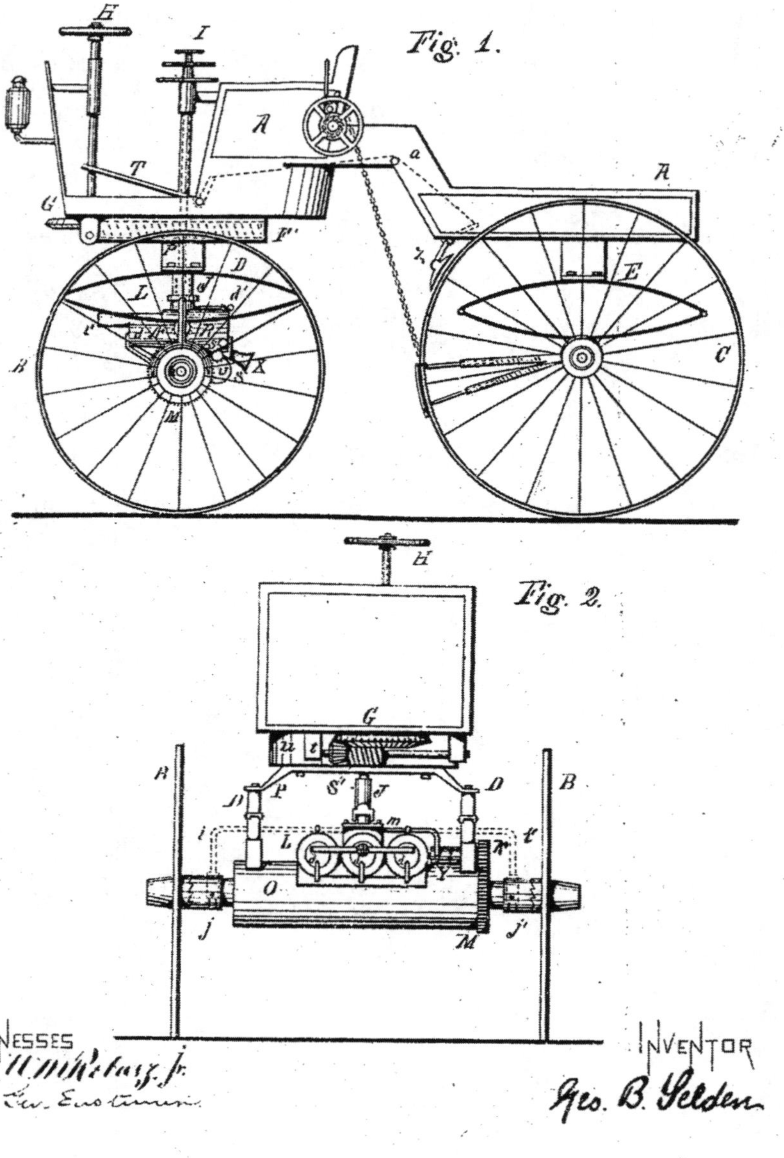

Fig. 1.

Fig. 2.

WITNESSES

INVENTOR

Geo. B. Selden

8 MORE DECISIONS. THE PATHWAY BEYOND GATE [8]

You've just completed a tutorial excursion on intellectual property fundamentals. Now it's time to establish and protect ownership of your idea by filing a patent application.

Looking back at our Map, you will see that you're at the crossroads just past [8]. There are various ways you can arrive at the next gate, Map point [15], which opens when your patent is filed. You could, I think foolishly, take the path along [9], on which you simply prepare your own patent application and file it yourself without a preliminary search. Or, you could take the path segment along [10] which leads you to a professional patent attorney or agent [14] who will do a search and file your application for you. Or, you can make a stop at [11] and study how to do a (DIY) patent search yourself. After you have done your DIY search, you could then either go to [14] for a professional search and filing, or you could go directly to gate [15] by filing the application yourself.

The recommended approach is to make a stop at [11] to learn how to do a DIY search. Once you have learned that, you can make your own search. Then if you wish, you can have a second, perhaps more thorough, search performed by a professional prior to filing the patent application. Just to summarize the recommended route, after passing through the gate at [8], make a side excursion to [11]; then take path [12] to point [14] for professional filing assistance. Here's why this approach is recommended.

When the USPTO receives a nonprovisional patent application, it's examined to determine into which category it falls and to judge if the invention is useful, novel, and non-obvious. Judgment on the latter two of these criteria is determined by searching the archives of all relevant, granted U.S. patents and published U.S. patent applications, as well as foreign patents. The search shows whether or not the claimed art has been anticipated by others. It's performed by a USPTO examiner who specializes in the technology sector into which the patent falls. When filing, you're obliged to advise the USPTO of any prior, close art that you're aware of. You're not actually required to perform any preliminary search beforehand, but it's advisable to do so.

Here's why. Unless you're extremely well informed on the existing art in your field, preliminary searching is likely to save money and a lot of hassling over the application's details. In processing the application, the USPTO examiner does a very thorough search. Your preliminary search allows you to tailor your application to avoid unnecessary and obvious conflicts with existing art that the USPTO would otherwise cite against you. It might even reveal that your art is not new at all, in which case you're sunk, but at least you avoid the whole application cost. It's better to find that out sooner rather than later.

No search is completely comprehensive, not even the search performed by the USPTO's examiner. There's such a vast amount of existing material, sometimes poorly catalogued, that relevant prior art is likely to be missed in even the best searches. Over eight million U.S. patents have been granted by the USPTO since its 1790 founding, and don't forget that foreign patents are also reviewed and can be cited against your application. Information archived within that mass of material can awkwardly surface after a new patent has been granted. When that happens, the patent might be re-examined, even after it has been granted. In light of the revealed information, its claims could be judged invalid. Taking that into consideration, it's best to be as thoughtful as possible in finding and revealing any close prior art before filing. That includes doing an adequate prior-art search.

The need for a rigorous prior art searching, and the care required in writing patent applications have both been made much more exacting with the recent implementation of the America Invents Act.[1]

Amongst other motives, it was enacted to bring U.S. patent regulations more closely into line with those of other industrialized nations. The changes, unfortunately, seem much more favorable to major corporations than to independent inventors. The AIA introduces increased opportunities on a number of grounds for patents to be challenged once they have been granted, and for their claims to be overturned. That puts an independent inventor in a very inferior, possibly indefensible position. Responding legally to such challenges would likely be very costly and time consuming. An aggressive attack could easily be a death sentence for an independent inventor's patent; he would probably not be able to withstand either the delay or the cost associated with the defense. It comes down to a question of who can best afford to be aggressive in determining a patent's ownership and validity. Even if a major corporation failed in their attempts to overturn an individual inventor's patent claims, the attempt, easily afforded by the corporation, could drastically weaken its vulnerable victim.

1 See www.uspto.gov : Global Impact of the AIA. See also Appendix I in this book.

The foregoing AIA comments serve to emphasize the need for very carefully prepared patent applications; however, the new review regulations are not quite as threatening as they might seem. Sound, credible evidence is required to initiate any of the aforementioned review processes. So, the fewer defects that exist in your patent's quality, the less you are at risk of being subject to review.

A more comprehensive discussion of the AIA and its implications is contained in Appendix I.

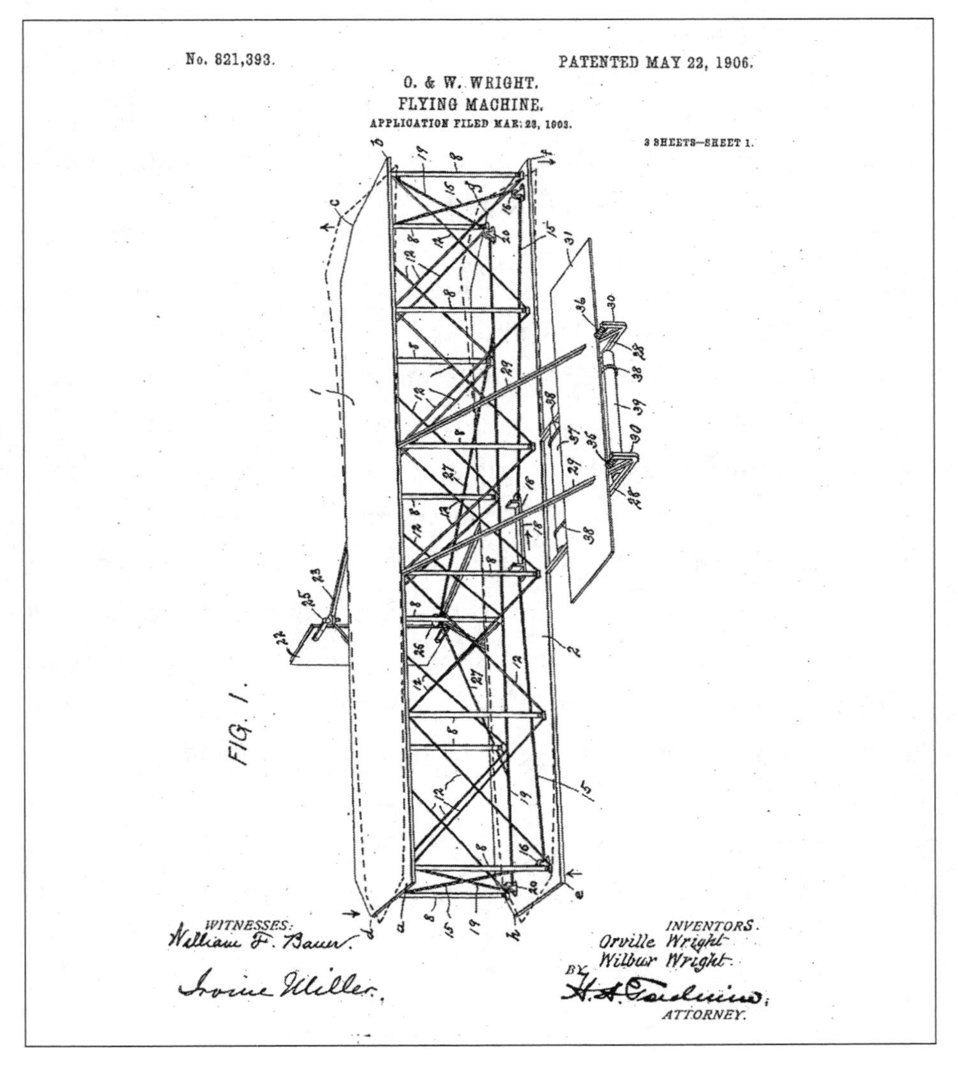

DO IT YOURSELF (DIY) PATENT SEARCH

A DIY preliminary search might not be as complete as professional searches, but it's a very good way for you to start. It will probably reveal some conflicts with existing art, and you'll inevitably learn a lot about potentially competitive products. Sometimes the DIY search will yield good ideas for your own invention, including uses for it you hadn't envisioned. You might also learn about products very similar to your own that haven't been successful in the marketplace. With a bit more research, including possibly contacting the unsuccessful products' inventors, you could probably learn why they failed. That would be extremely useful information. So you should view the DIY preliminary search as a valuable learning experience along the road to developing your product.

Searches can be performed using either hardcopy (printed material) or electronic databases. Both are available, but with the digitization of archived patent materials in recent years, the only sensible way for you to do your

DIY search is to access the electronic data. Computer (electronic) searches permit rapid database scanning for key words, phrases, or other indicators that flag relevant patents. You cannot do that with hardcopy databases, and the computer is right at your fingertips.

No matter how you search, you'll find some prior art that's very close to yours. Don't be discouraged by that. It's inevitable. If you don't find some existing close art, you're not doing a good search.

Almost no invention is entirely new. Americans often credit Henry Ford with inventing the automobile, for example, but of course he didn't. What he did was create a practical machine that could be economically built. He's the one who finally put it all together and made it work.

The fact is, lots of people had automobile ideas that preceded Ford's. Ford himself narrowly won a bitter patent dispute over G.B. Selden's prior art.[1] Selden never made a practical automobile. Ford did. Ford succeeded; Selden didn't.

It's also commonly thought that Thomas Edison invented the light bulb. Not so. Like Ford, he advanced an embryonic technology to practical use. But he, too, encountered prior art in the form of an earlier patent. In 1874 Canadians Henry Woodward and Matthew Evans patented a functioning glass light bulb that housed a carbon filament and nitrogen gas.[2] They did not have the funds to commercialize it, and a year later they sold the patent to Thomas Edison. The rest is well known history.

The prior art is out there; you have to find it and, if possible, distinguish your invention from it. When searching patents as a prelude to filing an application, you should look for several things. Primarily look for existing

1 U.S. Patent Number 549160 to Selden, 1895.

2 Canadian Patent Number 3738 to Henry Woodward & Matthew Evans, 1874; Also U.S. Patent 181,613.

art that could render your invention not-unique or that would make it obvious. Either of these would prohibit patenting your idea. The fact that your invention might be obvious in light of existing art would not necessarily keep you from manufacturing and selling it. It could be judged obvious with regard to existing art without actually infringing on that art's patent claims. Or it might be considered obvious in light of one or more expired patents. In either case, you might choose to go forward with the invention's development, protecting your position as well as possible with trade secrets.

To be thorough when doing the search, you'll want to search both issued patents and published patent applications.

As you make a first, cursory, overview of existing patents and applications, look at each one's title, drawings, abstract, and first claim. Usually, but not always, it's the first claim that gives the patent its broadest coverage. Keep a list of patents that seem disturbingly close to your invention, and later go back and review them in greater detail.

The USPTO offers a comprehensive online tutorial describing a methodology for searching patents for prior art. The tutorial can be found online at: **http://www.uspto.gov/web/offices/ac/ido/ptdl/CBT/.**

I recommend that you watch the tutorial. It lasts just over a half-hour, and will give you good insight into how U.S. patents are organized and how professionally conducted searches are done. The described method is best employed with the help of a librarian at a U.S. Patent and Trademark Resource Center (PTRC) Library. If you happen to be close to a PTRC library, you should go there and give it a try. It's thorough, but somewhat cumbersome.

The tutorial mentions some cautions about doing a key-word search using an electronic database. It notes that many words that might be entered into the key-word search have double meanings, such as "mouse," for instance. It also notes that patent titles and descriptions frequently use confusing terminology which can lead to missed prior art. Still, in my opinion key-word searching is the best way for an independent inventor to go about doing a preliminary DIY search, and it's certainly easy to do using Google's search capabilities. The USPTO has an agreement with Google to make U.S. patent and trademark data more readily available to the public free of charge. Many USPTO data products that are difficult to access through the USPTO's website are easily retrieved by going to: **www.google.com/patents.** That opens a search page in which you can enter any descriptive information you have regarding the technology you wish to search.

Here's an example: Suppose you have invented a new sort of collar for walking your pet snake. Go to the Google patent site, and enter "snake collar" into the search window. That will lead you to a list of patents having to do with snakes and collars, the first one of which is U.S. Utility Patent 6,499,999 in which the inventor describes his invention for recreational snake control (Figure 5).

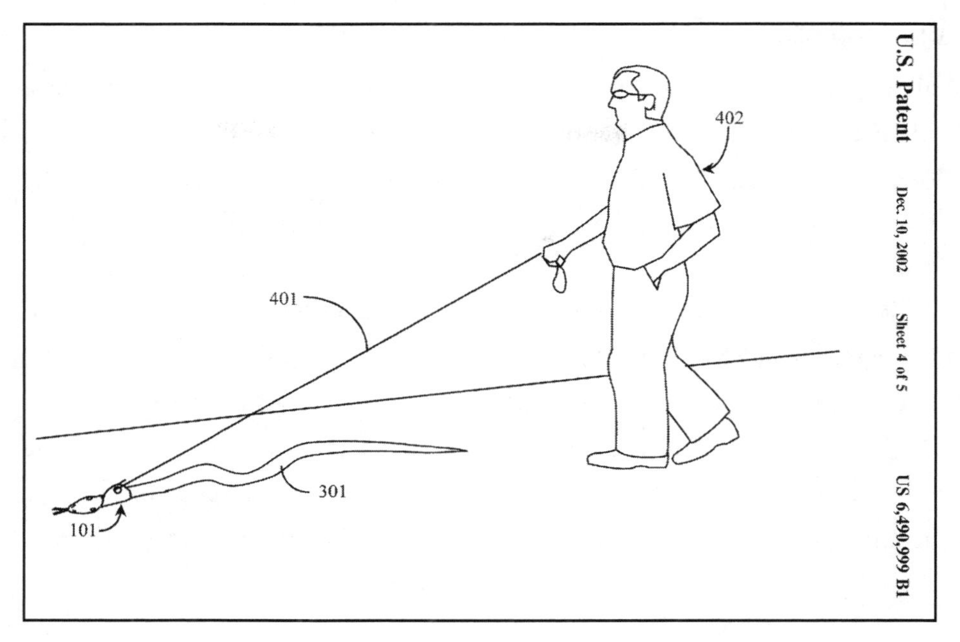

Figure 5

Click the '999 Snake Collar patent listed on the Google website. That will pull up another web page that lists the abstract of the Snake Collar patent, drawing images, the patent's detailed specification, claims, and a wealth of other information. Small drawings appear in the upper left hand area of the web page. If you click on them they will enlarge. On the upper right hand portion of the web page is a clickable icon "Download PDF." Clicking on that gives you the option of opening or saving the full image of the actual patent. What could be more convenient. In seconds the patent appears for you to print, save, or just peruse.

Scrolling further down on the Google web page you find a "CITATIONS" section which lists all of the close art that the USPTO patent examiner has cited in the Snake Collar patent's prosecution. It is this cited art that will have limited the scope of allowable claims on the Snake Collar invention. The cited patent numbers in the list are clickable, and if you click on any one of them the cited patent comes up with a web page just like the one we are talking about for the Snake Collar. You will notice that all of the cited patents have filing dates that precede that of the Snake Collar patent. That, of course, is because they contain art that was known *prior* to the examination of the Snake Collar patent application.

When you do a DIY search for prior art close to your own invention, you will want to look not only at the art cited against each of the close art patents you find, but also that which is cited against those patents listed in the CITATIONS section of the close art. To be thorough, it's best to go down a couple of layers into the cited art. That sounds like a huge task, but much of the cited art can be dismissed in seconds. The time it will take you to do your DIY preliminary search will be measured in hours, not days.

After the CITATIONS section there is another section titled "REFERENCES." This is a useful section that lists every patent in which the Snake Collar patent itself has been cited. It lets you look at work that has been done in your area of interest *after* the Snake Collar patent was filed.

Following the REFERENCES portion of the web page there is a "CLASSIFICATION" section listing into which ones the patent falls. Entries on the list are clickable. If you are doing strictly a key-word search the classifications might not help you very much; nevertheless, it is interesting to see into which classifications your invention falls. Again, it might give you some ideas for functions your invention could perform that you had not even thought about.

At the bottom of the Snake Collar Goggle page there is a section called "LEGAL EVENTS." It shows any legal actions affecting the patent's status including any assignments that have been made. Remember, if a patent's rights have been "assigned" to another entity by the inventor, the assignee owns the rights to the patent.

You will also see in the Snake Collar's LEGAL EVENTS section that the patent has expired due to non-payment of maintenance fees. If you were contemplating the manufacture of a similar snake collar, the fact that it's expired would be very interesting information; you would not have to worry about infringing the expired patent.

FIGURE 6: 1940's Patent Searching

In the end, when you have gone through the key-word search as described here, you will have done a reasonably comprehensive search for prior art close to yours. You should have a very good idea whether or not your invention is distinguishable enough over the prior art to merit filing a patent application. Not too many years ago the search routing just described could not even have been dreamt about. Until very late in the 20th century, searching patents for prior art was largely conducted by going through archived hard-copy material (Figure 6). That was an incredibly inefficient and time consuming task. I recall even in my own career having to spend days in a library doing it that way. The

USPTO website and its cooperative arrangement with Google have simplified the patent search process immensely. Whatever you enter into the Google patent search engine you will get the results drawn from the USPTO digitized archives going back to the inception of the U.S. Patent Office in 1790. It is truly remarkable.

The Official Gazette

Another way to quickly look at recently issued patents is to browse the Official Gazette, a weekly USPTO online publication of granted and expired patents. Each issue contains the first claim and a representative figure from every patent issued that week. That's just the information you want to look at for a quick search. You'll find the Official Gazette for the trailing 52 weeks on the USPTO's website. Unfortunately, Official Gazette patents are currently only searchable by patent number, classification, or patentee name, and not by keywords. And, they must be searched issue by issue, not globally. That's a nuisance, and severely limits the Official Gazette's usefulness.

Claims Infringement Analysis

One of the main reasons to do a patent search is to find existing art that might restrict the rights you hope to establish with your own patent.

Those rights, the "claims" you receive, will be limited by art that is close to yours. When you do find close existing art you must then assess to what extent it will affect your own patent claims. That is done by making an "*infringement analysis*" to determine if your invention trespasses on the rights of someone else's patent claims. There follows an example of how to do that.

The following mock infringement analysis serves as an example of how to evaluate to what extent your invention might infringe upon existing patents, or how your claims might be limited by them.

Let's say you've just invented the Figure 7 bottle/jar-opening device, and you want to do a preliminary search to judge whether or not your invention might infringe the claims of existing patented art. Never mind that your bottle/jar-opening device doesn't really seem very practical. It wouldn't be judged worthwhile by our earlier criteria, and if you study it carefully, you'll discover that it would be very difficult to manufacture. None of that matters. It wouldn't affect its patentability, and this is only an exercise.[3]

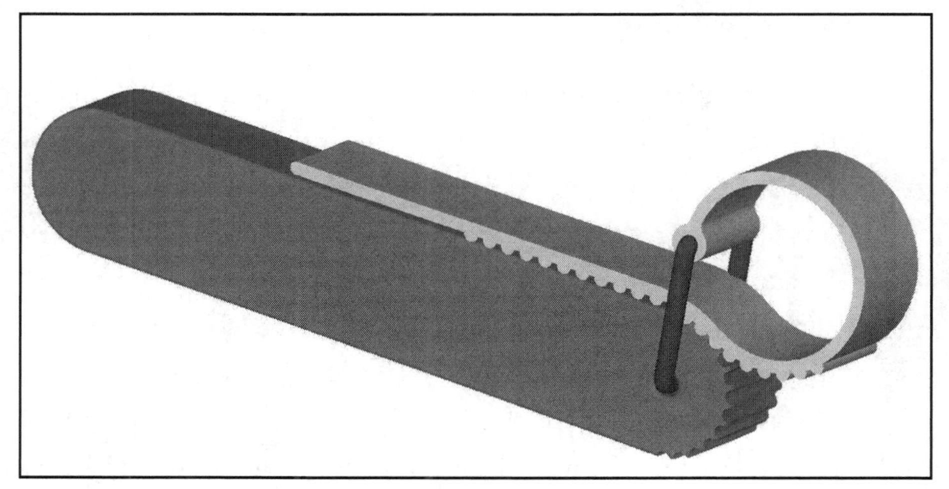

Figure 7

The bottle/jar-opening device you have in mind has a short rigid handle with a tough, flexible strap attached on one end to a buckle. The strap wraps around the bottle cap. The gears molded onto the strap and handle are an essential feature of your invention. They cooperate to tighten the strap and prevent it from slipping on the handle as the handle rotates in either the cap-openingcap

3 "Your" opener was just cooked up for this mock infringement analysis.

opening or, when turned over, in the cap-closing direction. They are intended to permit greater torque than could be obtained with smooth surfaces. Your invention is very similar to Halpin's in our earlier patent example, but let's assume you know nothing yet of Halpin 5,056,383 or any other openers of that sort.

As mentioned, there are a number of ways to proceed with the search. Here's a good one:

1. Go to: **www.google.com/patents.**
2. Enter "bottle, jar, strap, opener" in the search window. That will cause a list of patents to appear, and give you some more choices to make to narrow your search.
3. You can organize the list of patents by making choices from the menu bar that appears under the Google banner on the top of the page. Click on "Search tools," and another menu bar appears below the first one giving you a number of choices to make your search easier. The choices are pretty self-explanatory. Click on "Any patent type" and a drop down menu lets you choose the type of patent you want to search for. It's a utility patent you're planning to apply for in this exercise. It's not likely that an existing design patent would severely limit your eventual patent claims, whereas a utility patent might. So in this case you would sort them by patent type by choosing "Utility" patents. Back at the Search Tools menu there is another drop down menu titled "*Any Filing Status.*" You would search in turn on all "Issued Patents" and "Applications" by choosing them from the drop down menu. Again clicking the Search Tools menu there is a drop down menu titled "Sorted by Relevance." You can choose to sort the patents by date or relevance. In this search I chose relevance. When I entered the above search criteria more than one hundred hits came up.

4. Starting from what the search engine identified as "most relevant' at the top of the webpage and working your way downward, single-click on the number of any patent that seems interesting. The information on the selected patent will appear on a new page, just as it did for the "snake collar" patent we looked at earlier.

You'd go down the list of relevant patents, opening and examining ones that seem to describe art close to yours. If you do that, you'll open some with tantalizing titles. But, upon inspection, you'll probably see that most really aren't much like your invention.

One that somewhat resembles your bottle opener idea is that of our previous example: Halpin U.S. Patent 5,056,383. When I did the search, Halpin came up as the eighth one down on the list. You might study Halpin with some degree of apprehension, as it does indeed seem close to your idea. Although at first glance Halpin's patent looks like it might present an infringement problem, close study of his claims reveals that it does not. Here's how you can determine that. To organize the claims' study, it's useful to construct a claims analysis table such as that shown in Figure 8. It allows you to systematically analyze patent claims for possible infringement. The attributes of your device are compared to the claims of Halpin's patent in Figure 8 that lists all the language of Halpin's claims juxtaposed with corresponding comments on your invented opener. The language of Halpin's Claim 1 has been parsed into seven parts, and you will recognize that each part delineates an element of his invention.

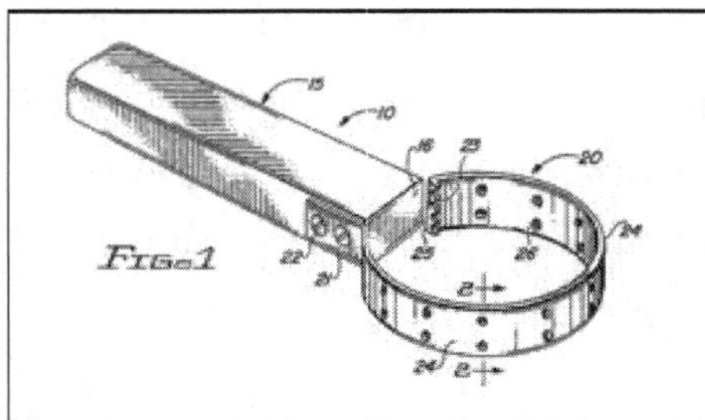	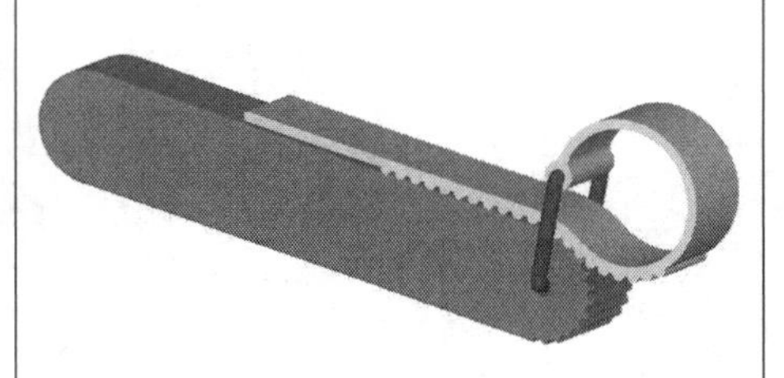
HALPIN U.S. PAT. 5,056,383 1ST, 2ND, AND 3RD CLAIMS:	**YOUR OPENER**
1. A tool for attaching and removing twist caps from containers	Yes, yours invention does that.
comprising a handle,	Yes, yours invention has that.
cap gripping means attached to one end of said handle	Yes, yours invention has that.
said cap gripping means further comprising a generally circular shaped band formed of relatively thin spring-like material,	No, yours is not spring-like. And is not shaped to be circular. Halpin relies on the shape and springiness to operate. Yours does not.
one end of which is attached to said handle with the opposite end thereof being unattached,	Yes, yours invention has that.
the surface of said band which is in contact with said cap being provided with gripping means,	No, yours has no special grip means. The tightening of y our flexible strap allows it to grip.
with the unattached end of said band being bent inwardly to provide additional gripping means.	No, the free end of your strap is not in contact with the cap.
2. The tool as in claim 1 wherein said band is made of spring steel	No, your invention doesn't have that.
and constitutes an arc of at least about 180°	Yes, yours is like that.
And wherein said unattached end of said band is bent inwardly at an angle of about 90°.	No, your invention doesn't have that.

3. The tool of claim 2 wherein said band constitutes an arc of about 360°	No, your invention doesn't have that.
and wherein said gripping means includes a series of spaced-apart inwardly facing projections.	No, your invention doesn't have that

Figure 8

As noted earlier, Halpin couldn't have been granted a patent with a first claim simply stated as, "A tool for attaching and removing twist caps from containers." That's too broad and would have resulted in conflicts with other patented inventions that pre-dated his. So he had to add the specific modifiers as shown in his first claim.

In order for your device to infringe on Halpin's first claim, it has to have all of the features specified by that claim. Your device has some of them, but it doesn't have them all; therefore, it would not infringe on that claim. Remember, the details of the claims very closely define the limits of the patent's scope.

Halpin's second and third claims are dependent on his first claim, and it's easily seen they pose no conflict with your invention as your invention does not have all of the elements that are claimed by Halpin.

In order for your device to be patentable over Halpin, it must be new art with respect to Halpin's invention. It seems that condition is met.

Secondly, your application's examiner must judge your invention to be not obvious in light of Halpin's teachings. In this case, it's not likely that it would be deemed obvious, as the basic functioning of yours is not taught by Halpin.

The foregoing example has shown how, by constructing an infringement table during your preliminary search, you could judge your device's possible patentability over one other existing patent: Halpin, '383.

Now you would proceed onward with the search to look for other close art. A very good way to go forward from this point is to go back to the Google patent site where, as you recall from an earlier discussion, all prior art cited by the examiner is listed.

The examiner has done a lot of work for you by citing all of the patents that limit the breadth of Halpin's claims.

As your invention is very close to Halpin's, these are most likely many of the same patents that will limit your claims, so you should look at them next. As you do, critique them in the same way you just critiqued the Halpin patent, making infringement-analysis charts as necessary.

The most recent utility patent cited against Halpin was Shaffer U.S. Patent 4,889,018, an excerpt of which is shown in Figure 9. Again, the invention is remarkably like yours, but the unique feature of Shaffer's device is the "recessed groove" enabling the opener to deal with lids having a reinforced rim. Yours doesn't have that, and therefore Shaffer's patent is not likely to be an impediment to patenting your device.

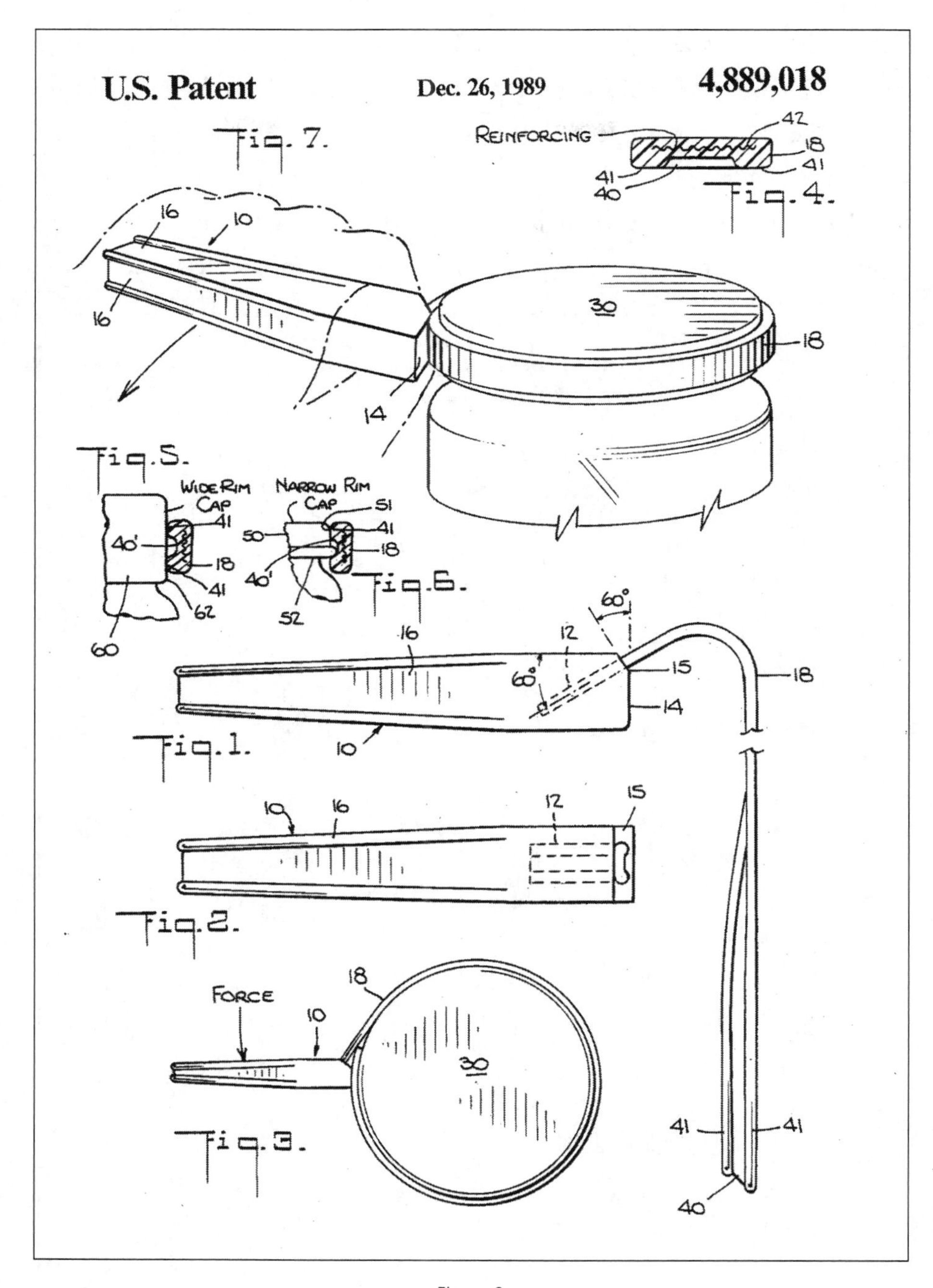
U.S. Patent
Dec. 26, 1989
4,889,018
Fig. 7.
Reinforcing
Fig. 4.
Fig. 5.
Wide Rim Cap
Narrow Rim Cap
Fig. 6.
Fig. 1.
Fig. 2.
Force
Fig. 3.

Figure 9

Continuing your search, you should look at the references cited in the Shaffer patent. One is Lewis U.S. Patent 2,771,802, part of which is shown in Figure 10. The Lewis patent is indeed close to your device. Still, it doesn't incorporate your strap handle's geared feature. So your invention would still be patentable over Lewis; however, even though Lewis's patent was granted in 1956 and has expired, his patent would still possibly limit the scope of your claims.

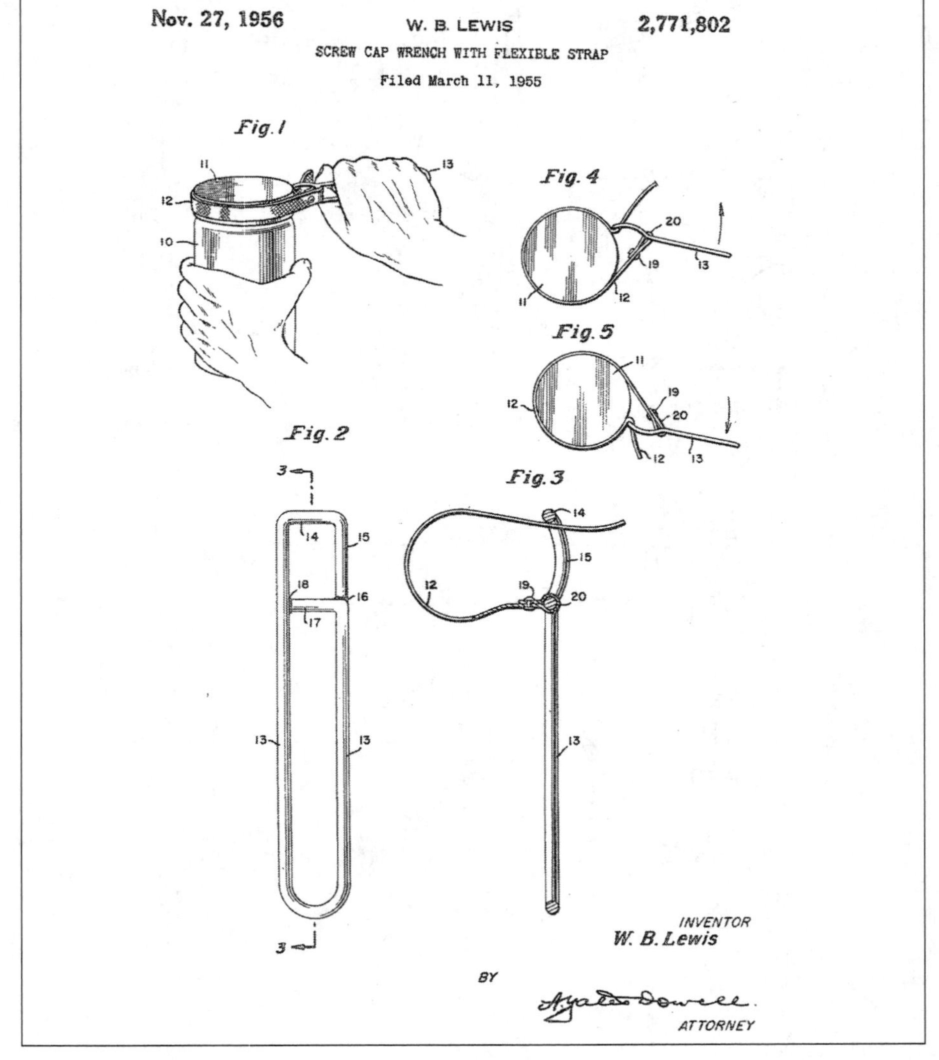

Figure 10

As you examine the close-art patents for infringement you should also look for disclosed content beyond that which is claimed. Everything within the patent including figures, descriptions, and so forth constitute openly published material. Except for what is actually claimed, it's in the public domain, meaning that it is fully available for public use and could be employed in other inventions. On the other hand, the fact that it is publically disclosed means that it cannot be claimed again in another patent; therefore, it might limit your claims even though it does not present an infringement problem.

As mentioned earlier, if you are examining patents on the Google patent site, there's a section titled "Legal Events." Check it out as you go along to see if the patents have expired due to non-payment of maintenance fees or any other reason. If they have, then everything disclosed in the patent, including the claims, is in the public domain and available for use.

By now you have the idea of how to proceed with an electronic DIY preliminary search and what to look for.

The foregoing exercise was intended to show the way. In an actual search, you would painstakingly track down all interesting leads.

Summarizing this section, DIY searching is recommended for a number of reasons including the following:

- It increases your familiarity with existing art
- It often reveals unanticipated applications for your invention
- It can provide valuable insight into what potential competitors are doing
- It can reveal information about similar products that might have failed in the marketplace

- It sometimes suggests invention improvements
- It aids in preparing your own patent application
- It's inexpensive, and if it reveals unavoidable conflicts with prior art, it saves you from more costly procedures.

I have spoken with inventors who have done, or have had done for them, patent searches to find art that might keep them from getting a meaningful patent on their own invention. They found art that appeared to be discouragingly close to theirs, and scrapped their projects. What they did not do, nor have done for them, was a proper claims infringement analysis. I found in more than one case that the existing art did not actually impact their invention at all, and they were able to successfully resurrect their projects. The point is: do not be intimidated by close art that superficially appear to conflict with yours. Remember, *every* word in the claims of a close-art patent limits the boundaries of that patent owner's intellectual property.

Please go to the Quiz on page 259.
You should now be able to respond correctly to quiz items 81 through 91.

PROFESSIONAL PATENT SEARCHES

Let's have a look back now at our Figure 1 Map to see where we go from here. If you have followed the recommended route so far, you will have just learned how to do a DIY search, and will have seen how to judge the chances that your invention might infringe on existing art. Assuming you've had a successful DIY search, from your location at point [11] on the Map you can either take path segment [13] which goes directly to filing your own patent, or you can take path [12] to get professional help at [14]. The agent or attorney at [14] could do another patent search for you and then file, or he could file without a search.

Unless you have extensive personal search experience or are extremely well informed in your field's current technology, I recommended you have a professional search done to supplement the one you have done yourself. Be sure to share your DIY search results with your patent professional; they could be helpful to him.

If you hire a patent law firm to do a search, they in all likelihood will, in turn, hire a search firm to perform the task. Most legal professionals have search firms who regularly do their work and who are known and trusted by them. Your agent or attorney will mark up the search firm's bill by some amount, typically doubling it. His markup covers facilitating the search and rendering his professional opinion of the results. A search firm will provide only the search results, not an opinion regarding them; so whether you go through a law firm or search firm you'll get the same search results. The main differences are that the attorney or agent will provide an opinion of the results...and you'll pay more. Another difference is that unlike professional search firms, your patent attorney or agent is registered with the USPTO, and you probably know something about him. His familiarity with the preliminary search results will help him tailor your patent application to avoid discovered prior-art interference. That's something a search firm cannot do.

You can disclose your ideas to your attorney or agent without worrying about him violating their confidentiality. The firm he will employ to actually do the search should be bound by him to keep your ideas confidential. You can check with him to be sure that's the case. Search firms themselves are not registered by the USPTO, and until the inventor has some experience with them or has a trusted reference, it's better to pay the law firm's supplemental charges.

There are unscrupulous invention marketers who disguise themselves as search firms; so beware. If you bypass your attorney or agent and go to a search firm directly, be very sure the search firm is legitimate. Some fake ones are very clever at misrepresenting themselves. You cannot be too careful.

Before ordering a search from your attorney or agent, ask him which firm does his searching. Ask him how much the search firm will charge and how much he will mark up the price. If his answers are not satisfactory, go somewhere else. As a very rough estimate, at the time of this writing, an uncomplicated utility-patent search with opinion costs around $1,500.

A professional conducts the search in pretty much the same way as that described for the DIY search and uses the same resources plus some other search tools not available to the general public. Many search firms have the added advantage that they are located near the USPTO's search rooms and so have everything practically at their fingertips, including the expanded electronic search routines available there. Of course, they also have the advantage that they do it every day and really know their stuff.

The U.S. Patent and Trademark Office

Anyone fortunate enough to be near to the U.S. Patent and Trademark office in Alexandria, Virginia can go there to conduct both computer and hardcopy patent searches. The personnel there will gladly help you with your search. The Patent and Trademark Office is located at:

Public Search Facility
Madison East, 1st Floor
600 Dulany St.
Alexandria, VA 22313
571-272-3275

When you go, be sure to have some sort of official picture identification with you to get through the security barrier. A current driver's license will do. Once in the search area, you'll have to get a permit if you wish to use the computer search facility.

If you ever have the chance, you should go to the U.S. Patent Office. It's extremely interesting. The personnel are exceptionally helpful, and the search facilities are superb. Be sure not to miss the USPTO Museum and the National

Inventors Hall of Fame located just across the atrium from the search facility. You can go there without obtaining any entry pass, and you will learn a lot of interesting patent history. Tours can be arranged in advance.

From 1790 through 1880, working models were required as part of U.S. Patent applications. Despite devastating fires that destroyed the patent office in 1836 and again in 1877, many models have survived the intervening years, including the Civil War period. A few of these very intricate scale models can be seen in the USPTO museum. Others are in the Smithsonian Museums, and still more are scattered in private collections. The Rothschild Petersen Patent Model Museum in Cazenovia, New York, houses nearly 4,000 patent models and related documents spanning America's Industrial Revolution. It is the largest private collection of patent models in the country. Some of the models can be seen on the Rothschild Petersen Museum's website at: **http://www.patentmodel.org/.**

Many of the models themselves are wonderful, ingenious works of art and craftsmanship. The website also gives information on travelling exhibits, and other permanent exhibits of patent models.

PATENT FILING

Once your patent searching is finished, you're ready to go on to file either a provisional or nonprovisional patent application. The discussion in this section refers only to U.S. patents. Foreign filing will be touched upon later in the text.

Any inventor is permitted to file his own patent application directly with the USPTO. In the *Pro se – Pro bono* section of the USPTO's Inventor Resources webpage you studied earlier you will have found instructions for self-filing. There are also some self-help books available in most bookstores that guide you through the self-filing process. But doing it yourself really isn't a very good idea. If your invention's worth patenting, it's worth paying for a professionally written application if you possibly can. You'll have enough challenges along the way without starting off with a weak patent. Competitors will be looking for any way possible to get around your protection, particularly if your product is very successful.

I stress, though, that you certainly may file your patent application yourself. And, you might have to if your financial circumstances do not allow you to get professional help. But you're in for a lot of work, and most likely will not get patent protection that's as strong as you would get working with a professional. I am currently mentoring some inventors who are without funds, and are doing their own filing. Even though they find the USPTO personnel very helpful, they are investing a lot of time and frustration in the process. Avoid that unless you simply have no alternative.

Many self-help books purport to guide a beginning inventor through the self-filing process.[1] This text does not; instead, it offers advice on how to get the best possible patent when working with a professional attorney or agent. Even if you decide to have a professional file your patent application, you still have an important role to play in the process.

Your first step is to find a good attorney or agent. The USPTO publishes a registry arranged by states, cities, and foreign countries of individuals authorized to file patent applications on your behalf. It's available on their website[2]. It lists *patent attorneys* and *patent agents.* Both attorneys and agents can prepare patent applications and conduct their prosecution in the USPTO. Patent agents, however, cannot practice law. For example, if a patent agent's state considers drafting contracts as practicing law, he couldn't draw up a contract relating to a patent, such as an assignment or a license. He couldn't represent you in any litigation, and he shouldn't offer you any legal advice. Registered patent attorneys, on the other hand, are permitted to do all of that.

1 Be very careful using self help books on patent filing. Due to recent U.S. patent regulation changes all that I have seen are out of date. Check their publication dates. Major changes occurred in early 2013 with the America Invents Act. Prior publications will not reflect them.

2 Go to http://www.uspto.gov/inventors/index.jsp. Then click on "State Resources."

Persons not on the registry aren't legally permitted to represent you before the USPTO. Appearing on the registry doesn't constitute a USPTO recommendation; it simply means the registrants met the listing criteria.

Finding a competent representative isn't always easy. When you choose an agent or attorney bear in mind you're probably entering into a long-term relationship. You'll make an expensive investment educating him on your technology. Patents often require professional attention for many years, and it's not convenient to change representatives once you've brought one up to speed on your technology. Choose very carefully. Look for someone experienced in your field of interest. If you invent mechanical devices, for example, you'd seek an attorney or agent who has a strong mechanical engineering background. One with an undergraduate degree in, say, chemistry might not be the best choice.

To find a representative, go to the USPTO's directory. Find the names and addresses of registered representatives in your area. Pick out a few that seem promising and interview them. Try to find one you're comfortable being around.

Here are some things you might ask candidates:

- How long does it typically take you to prepare and file a patent application?
- Do you write the patents, or is some part of it done by an assistant?
- What are your hourly rates? How do you bill phone consultations?
- Who does your searches? How much do you mark them up?
- Will you accept patent drawings submitted by me? Are your patent drawings machine-drawn or hand-drawn? Is your computer drafting program compatible with my design software?

- Has any member of your firm ever been suspended from practice before the USPTO?
- My work is mostly on mechanical (or whatever) devices. How will your background be adequate to file patents in that technology area?
- Ask to see issued patents that the interviewee has prepared himself, not just ones prepared by his firm. (You can check them out on the USPTO's site.) Some patent attorneys are very good at practicing patent law but less stellar at writing patents.

Remember, the attorney or agent you choose will work for you, not the other way around. State firmly what you want. You can always go somewhere else early in the relationship. Once you're well into the process, however, it's costly and time consuming to change.

When employing a patent attorney or agent, you will execute a power of attorney or authorization of agent, which must be filed in the USPTO. It's usually part of the patent application. Once your representative has been appointed, the USPTO does not communicate with you directly but conducts all correspondence with him. However, you're still free to contact the USPTO yourself concerning the application's status. Contact information is contained on the USPTO's website. You may remove the attorney or agent any time by revoking his power of attorney or authorization.

I suggest you engage a patent agent associated with a full-service patent law firm to prepare your application. In my own experience, I've found agents to be better than attorneys at writing applications, and they are less expensive. Being part of a patent law firm allows the agent to work intimately with a patent attorney when the need arises. Furthermore, the law firm you choose should have foreign correspondents with whom they regularly work to facilitate filing abroad when that time comes.

It's essential for you to know patent contents and format, both to understand how to search patents and to assist in filing them. You should view the patent application as a joint project between you and your patent attorney or agent. And, by the way, you shouldn't just turn the whole works over to him and let him take it from there. You must stay involved to get the best result.

Here's one excellent way to do that. Once you're familiar with the application's expected format and content, do your level best to prepare it yourself. Really work at getting it just right. Then, give what you've prepared to your attorney or agent. Let him take the lead in completing it. Most agents and attorneys welcome this sort of inventor participation. It makes it easier for them to understand the invention and what the inventor sees as its most important features. This joint approach's end product will be superior to that achieved by either party working alone. You should try to help your attorney or agent as much as possible without impeding his work.

Foreign Filing:

So far, we have only considered filing U.S. patent applications. It is often desirable to also obtain intellectual property protection in foreign countries. There are several ways to approach foreign filing. The easiest and most cost-effective way to begin is to file a single Patent Cooperation Treaty (PCT) "international" patent application, which gives the inventor interim protection in the 148 countries which now participate in the treaty.[3] They include all of the major industrialized countries in the world.

3 If you wish to learn more about it, go to: https://uspto.connectsolutions.com/pph/. This is the best explanation of the PCT I could locate. It's more than the independent inventor would normally need to know. You will also find helpful PCT information at www.uspto.gov.

Foreign filing is not discussed here to prepare you do it yourself; it's simply to make you aware of the PCT filing option. And as you will see, the PCT application opens the door to a strategy that allows you to move some of the U.S. patent filing expenses a little further down the road. Here's how that works.

Within a 12-month period after the effective filing date (EFD) of your U.S. provisional patent application you are permitted to file a PCT application in lieu of a nonprovisional U.S. patent application. The PCT application usually costs much less than a nonprovisional U.S. patent application and it allows you up to another 18 months from the PCT's filing date to file individual applications in the nations in which you are seeking protection. One of those nations can be the United States.

Said in other words, a PCT application allows you up to 30 months from the filing date of your U.S. provisional patent application to decide whether or not to file a non-provisional U.S. patent application, or to file individual applications in other countries where you want protection.

Exercising the above PCT options, you can stretch out patent filing fees in the following way:

1. First, file a U.S. provisional patent application on your invention not more than 12 months after it is first publicly disclosed. (Cost about $300 to $500, and it buys you up to a year's protection.)

2. Just before the passage of 12 months from filing your U.S. provisional patent application file a PCT patent application. (Cost about $3000, and it buys you another 18 months of protection.)

3. Just before the passage of 18 months from filing your PCT application, file a non-provisional U.S. patent application, and

corresponding applications in any other countries in which you seek patent protection. (Cost around $10,000 to $15,000 for a U.S. utility application, and additional costs for each other country in which you want protection.)

Filing in each foreign country separately is very expensive due to the facts that the application must be translated into the country's language, and examined according to the country's criteria; additional filing fees must be paid, and so on. The PCT application is written in the single language of the nation in which it is filed, in your case probably the United States, and you pay one set of fees. So, the PCT application allows you to defer those individual national filing expenses, including those for the United States, for up to 30 months from the filing date of your U.S. provisional patent application.

A PCT application also gives you more time to decide in which nations you really need protection. It allows you more time to see if, and where, you have a major market for your product.[4]Finally, it gives you a better chance to see if your product merits the expense of filing nonprovisional patent applications in the U.S. and other nations.

There are many facets to PCT international filing. To learn more about the subject you can go to: **http://www.wipo.int/pct/.**

There you will find a very comprehensive explanation of PCT international patent filing.

I recommend that when you get around to the point of PCT or other foreign filing you discuss all available options with your attorney or agent.

4 There are some other foreign filing possibilities wherein one application is valid in several countries. The European Patent Office (EPO) and Japanese Patent Office (JPO) offer some blanket coverage. Like the USPTO's regulations, the EPO's and JPO's are in a state of flux while presently undergoing some improvements.

No matter which of the pathways you take to get your patent application filed, once that's done you're faced with an exciting new set of challenges and opportunities. You have established ownership of your own intellectual property. You now have something tangible to lease, sell, or further develop. The next leg of our journey will take us past Map point [17] where we'll explore the various decisions that await you further down the path. Before that, though, I want to caution you about some undesirable characters that will be pursuing you for most of the rest of the trip. They are unscrupulous invention marketing and development firms.

Please go to the Quiz on page 259.

You should now be able to respond correctly to quiz items 92 through 106.

12 FRAUDULENT MARKETING AND DEVELOPMENT FIRMS

Once your patent application is published you will begin to feel very popular. You'll start to receive glowing letters and slick brochures from companies who think yours is the best invention since copper wire; they're eager to help you develop and market it. You might be invited to fly across country just to meet them. Don't be fooled by them; many, probably most, are just slick scam artists.

There are lots of fraudulent predators disguised as legitimate invention marketing and development firms. You have probably seen some of their ads on television. They claim to provide inventor services such as patent filing, licensing, marketing, research, financing, manufacturing, product distribution, and technical support. They prey on the vulnerability of inexperienced inventors who have patented or in some cases even unpatented ideas. Some of their advertisements even urge you to reveal your unpatented invention to them in confidence, and they'll take it from there… for some serious up-front money.

In a March 25, 1999 address before the United States House of Representatives, Q. Todd Dickinson, then acting Commissioner of Patents and Trademarks, and Acting Secretary of Commerce, stated: "One of the greatest threats to the integrity of our process is the proliferation of so-called invention promotion or marketing organizations. There can be little doubt that the scandalous invention marketing schemes are among the greatest problems faced by the independent inventor and very small business concerns. Raking in more than $200,000,000 each year, most often from those who can least afford it, these fraudulent firms do more than simply take the inventors' money; they rob them of their hopes and dreams. The damage to America wrought by these firms goes to the heart of our free enterprise system, and serves to depress and discourage one of our most unique sources of new ideas."

Each year many thousands of novices fall victim to the false advertising of these firms.

The U.S. Federal Trade Commission has prepared some tips for spotting such fraudulent firms and notes that many falsely claim they can turn almost any idea into cash.[1] Here are some of the frequently employed come-ons they found and how to realistically view them:

"We think your idea has great market potential." Few ideas, however good, become commercially successful. If a company fails to disclose that investing in your idea is a high-risk venture and that most ideas never make any money, beware.

"Our company has licensed a lot of invention ideas successfully." Wonderful. If a company tells you it has a good track record, ask for a list of its successful clients. Confirm that these clients have had commercial

1 Please see: http://www.uspto.gov/web/offices/com/iip/documents/scamprevent.pdf

success. If the company refuses to give you a list, it probably means they don't have any.

"You need to hurry and patent your idea before someone else does." Be wary of high-pressure sales tactics. Although some patents are valuable, simply patenting your idea does *not* mean you will ever make any money from it. Some firms purport to file inventor's patents, when in fact they file only provisional patent applications or nothing at all.

"Congratulations!. We've done a patent search on your idea, and we have some great news. There's nothing like it out there." Many fraudulent firms claim to perform patent searches. If they do so at all, the searches usually are incomplete, conducted in the wrong category, or unaccompanied by a registered attorney's opinion of the results. Because unscrupulous firms promote virtually any idea or invention without regard to its patentability, they might market an idea for which someone else already has a valid patent. In the highly unlikely event that your invention is successfully promoted, you could be the defendant in a patent infringement suit.

"Our research department, engineers, and patent attorneys have evaluated your idea. We definitely want to move forward." This is a fraud's standard sales pitch.

"Our company has evaluated your idea and now wants to prepare a more in-depth research report. It'll be several hundred dollars." If the company's initial evaluation is "positive," ask why the company isn't willing to cover your idea's additional researching costs.

"Our company makes most of its money from the royalties it gets from licensing its clients' ideas. Of course, we need some money from you before we get started." If a firm tells you this but asks you to pay an up-front fee or credit payments, ask why they're not willing to help you on a contingency basis. Unscrupulous firms make almost all their money from advance fees.

Here's some advice: If the marketing firm is overly enthusiastic about your invention's potential, but wants an up-front fee to help you get it commercialized, walk away quickly. Legitimate agents don't rely on large advance fees. They work on a contingency basis, not making any money unless you do. So, they're very selective about which inventions they agree to market. Since very few inventions ever make any money, they have to bet their time and money on likely winners.

The American Inventors Protection Act of 1999 gives you certain rights when dealing with promoters.[2] Before an invention promoter can enter into a contract with you it has the *duty* to disclose the following information about its business's practices during the past five years:

- How many inventions it has evaluated
- How many of those inventions got positive or negative evaluations
- Its total number of customers
- How many of those customers received a net profit from the promoter's services
- How many of those customers have licensed their inventions due to the promoter's services
- The names and addresses of all invention promotion companies it's been affiliated with over the past ten years.

You must insist that the promoter gives you this information; get it signed and in writing. It can help you determine how selective the firm has been in deciding which inventions to promote and how successful that promotion has

2 To order a copy of the American Inventors Protection Act, visit www.uspto.gov/web/offices/com/speeches/s1948gb1.pdf or call the USPTO.

been. Use this information to determine whether the company you're considering doing business with has been subject to complaints or legal action. If the company's reluctant to give you this information signed and in writing, they've failed the integrity test; they are not legitimate.

The promoters from whom I've received solicitations have politely declined to provide me the requested information, but they continue to send me solicitations. Or, they claimed to provide some other services such as patent searching, thereby slipping out of the disclosure duty of marketing and promotion firms.

Typically a fraudulent invention-promotion firm will entice the victim into a contract which will include design work, prototype manufacture, promotional brochures, models or video, licensing agreements, and so on. They might propose a two-step program beginning with a several hundred-dollar evaluation. That doesn't sound like much money, and it entices the victim in. The evaluation will be very positive, alluding to the prospect of great success, and further luring the victim into a much more expensive second phase. Some months after they have taken the victim's money, the agent with whom the victim worked will be said to have left the company. A string of excuses will follow as to why no real work has been done, and finally no one will talk to the victim any more. That bit of agony can cost from $5,000 to $25,000 and leave the inventor wasted.

I have heard of money being recovered in the case where the inventor has been defrauded by a bogus marketing firm through the U.S. mail. That's a federal crime. But unfortunately, in most cases once a fraudulent organization has your money it's pretty surely gone forever.

For additional information on promoters, you can call the U.S. Patent and Trademark Office, the Better Business Bureau, the Consumer Protection Agency, and the Attorney General in your state or city and in the state or city where the promoter's company resides. Your local chamber of commerce and business/manufacturing organizations might also be helpful.

If you have complaint with a promoter, register it with the USPTO. They'll forward your complaint to the promoter and publish its response online.[3] As an informative exercise, you should look at the list of registered complaints at: **www.uspto.gov/inventors/scam_prevention/complaints/index.jsp.**

Visiting the above USPTO webpage will allow you to read some of the unpleasant experiences that other inventors have had, and see which companies have bilked them. Inventing is a noble and patriotic endeavor; unfortunately, inventors must constantly guard themselves against these unscrupulous imposters.

Having said all that about phony marketing and promotion firms, it's still a fact that independent inventors occasionally do need outside help to develop and market their ideas. So where should they turn? Finding reliable, honest assistance can be a real problem. One way to approach it is by contacting legitimate inventor-assistance groups for guidance. The U.S. Department of Energy (DOE) published a list of such organizations in their Inventor Assistance Source Directory.[4] All of the organizations listed are non-profit. The directory lists state, federal, and private organizations in a matrix that gives each name, address, provided services, and contact information. Through it you can locate inventor assistance groups in your area that provide various types of help, including market and technical assessments, investments, and other funding and planning assistance. It also gives you a way to identify and interact with other local inventors. The DOE does not endorse the listed groups. You need to check them out yourself before using them, but the directory is a good place to start.

3 You must submit it on the form "Complaint Regarding Invention Promoter," which you can do at http://www.uspto.gov/web/forms/2048a.pdf.

4 You can download it from http://www.greenbiz.com/toolbox/tools_third.cfm?LinkAdvID=6942. Although published in 2000, it's still a valuable rescource.

CASHING IN ON YOUR INVENTION

There's no rule that says you cannot somehow cash in on your intellectual property anywhere along the way, but there are some points where the prospect is particularly attractive. These are generally where the invention's progress has just achieved a notable milestone. Aside from identifying your fundamentally good idea, the first significant milestone is establishing its ownership through a filed patent application. At that point, you own something tangible.

Recall the earlier analogy between a patent and a land deed. Now you have a deed to your intellectual property, or at least a pending deed. That's something you can sell or lease. There are now a few common ways to acquire funding and/or get some cash in your pocket.

We'll spend a considerable amount of time at this first logical funding opportunity to examine the possible business alternatives. In order to make intelligent choices, it's necessary to look at all the various possibilities in some detail.

Here are some of the options you might have at this point:

- **Assigning** (selling) some or all of your ownership
- **Licensing** (leasing) rights to it
- **Partnering**
- **Starting your own invention-based business**

At various times in my career, each of the above ways to cash in on my inventions has been appropriate. I'll relate some of my own experiences as we go along.

As you contemplate your own options, note that each one has potentially serious pitfalls. Except for the outright sale of all interest in your invention, every other choice leaves you in a long-term, intimate business relationship with another party. Thinking of it more or less as a marriage will keep it in perspective. How long and intimately you will be involved in that marriage depends somewhat on the nature of your invention. Here are a couple of extreme cases to demonstrate what I mean.

Let's consider a very simple child's animated toy, a typical example of an uncomplicated invention. Once the idea's concept has been designed, it's ready to be made and sold. There's little more the inventor can add to its development. The challenges are to manufacture and package it economically, and to market it effectively. All of the above options for cashing in are available to the toy's inventor. If he wants out early he can exit the game via an assignment or licensing agreement and still get a good payout on his ideas with little or no ongoing personal involvement.

On the other end of the spectrum, consider a complicated piece of leading-edge fiber-optic communications equipment. The inventor has probably built a mock-up and demonstrated it in the laboratory. But it still has to be perfected technically. Performance evaluation and integration protocols must be developed. Test and prototype-manufacturing equipment must be assembled. Technicians have to be trained. Potential users will have to upgrade their systems to utilize it, and much, much more. A lot of time, money, and effort are still required to get the product on the market.

Even though in principle the fiber-optic's inventor could pursue the options to assign or license the rights to his device, he has a different future ahead of him than does the toy's creator. Because his invention's so complex, it's unlikely that there's a quick exit for him no matter which option he chooses. There's just too much product development left to be done, and he's probably going to have to be involved to make that happen. He'll most likely be in the marriage for a long time.

The next few sections explore each of the first three abovementioned cashing-in options. There's an in depth discussion of the fourth option, starting your own business, later in the book.

Patent Assignment

An assignment is the outright transfer of an issued or pending patent's ownership. Generally an assignment is not for a fixed, limited time; it's for the lifetime of the patent. *Full assignment* is the complete transfer of all interest in the patent. *Partial assignment* transfers only a portion of the patent's ownership.

In many cases, even when you have assigned all or part of your intellectual property rights, you will be contractually bound to stay on to further the development of your invention. If you do that, don't undervalue your expert role in the continued development. Be sure you are guaranteed adequate compensation for your ongoing work. And, be sure to set a reasonable limit to the length of time you are obliged to stay on. One to two years is not uncommon.

Sometimes when an inventor assigns rights to his invention, particularly if he stays to further his invention's development, he will be asked to sign a non-compete agreement with the assignee for some period into the future. Be extremely careful when entering into such an agreement. The laws concerning non-compete agreements vary from state to state. Get an attorney from the state having jurisdiction of the agreement to review and explain it thoroughly to you before signing. If possible, limit the agreement's rights to the assignee, and not to his successors. In other words, the non-compete agreement should not be transferrable by the assignee. Vague covenants in such agreements can leave them open for future unfavorable interpretation by the assignee.

Full Assignment

You might not have any choice whether or not to assign your intellectual property. If you're under an intellectual property agreement to your employer, the agreement probably requires that you fully assign all patent rights to your company and that you assist in prosecuting any future activity associated with your patent. As an example, you would normally be required to assist your employer in defending the intellectual property against infringers. When you assign all of your rights, you remain the inventor; nothing can change that, but you're no longer the owner.

If your intellectual property's ownership is unencumbered, full assignment is one straightforward way to get money for it. In return for agreed upon compensation, you could sell the assignee all of your rights and go on your way.

If your invention requires further market research and/or development or if your patent's still pending, you would probably not maximize your return by selling early. The assignee would still see considerable risk and expense ahead to define, produce, and market the product. And then he would have to wait to see if it's successful. You could get some return for your property by selling it at this point, but the return will likely be greater if you wait until your program is further advanced.

On the other hand, selling your entire interest early, even on a complex invention, might be advantageous. Maybe you're mentally burnt out on it. Or, perhaps you've lost interest or have other more fruitful projects in the works and find yourself without any more resources to devote to it. In such cases, you might settle for diminished assignment returns just to get on with other things.

Partial Assignment

Of course, just as with real property, you don't have to sell all of your interest. You can sell part of it. In return for something, usually cash, you can assign part ownership to another person or business. A full assignment typically sets you free, whereas a partial assignment probably doesn't.

Suppose as a partial-assignment example you estimate that the value of your invention would be tripled if you had $20,000 to construct some presentable prototypes and literature. You have no money to do so. You have a friend who offers you the $20,000 for part ownership in your invention. After negotiation, you might agree to take the money and assign 10 percent of the patent's ownership to him. The assignment agreement should clearly specify what he does and doesn't get for his investment. Clearly, a partial assignment agreement is somewhat more complicated than a full assignment one, because the assignment contract can be whatever the signatories agree to.

The partial assignment puts you in business with your new co-owner. As noted earlier, U.S. patent law gives joint patent owners equal rights to exercise the co-owned patent. That means unless it's otherwise contractually specified, your 10 percent partner has as much right to exercise the patent rights as you do. That's probably not what you had in mind.

You can avoid that by making your assignment agreement state exactly what the new part-owner's rights and limitations are. When you partially assign, carefully consider the following things. Be sure they, and all your other issues, are clearly specified in a written agreement.

- What respective patent exercise rights will you and the co-owner have?
- Will your new co-owner be a working partner or just an investor?
- If he's a working partner, what are his duties and compensation and what are yours? What happens if one of you doesn't do his part either voluntarily or through death or incapacitation?
- How will patent revenues be shared?
- How will expenses be apportioned?
- How will product liabilities be shared?
- Who will have ultimate decision-making authority? How will the decision process take place?
- How will differences of opinion be resolved?

- How will rights to exercise subsequently patented improvements be shared?
- How will the co-owners share in expenses of defending the jointly owned patent(s) against infringers or against those who might claim infringement?
- What happens if the assigned patent is pending and is never issued, or if after it's issued its claims are subsequently overturned through re-examination?
- What rights will the co-owners have to sell their interests? Will there be any rights of first refusal for the other co-owner(s)?
- How will rights and duties transfer in case one of you dies?

Because there are so many things to consider when entering into a partial assignment agreement, you should have an intellectual-property attorney assist you with the assignment document, just as you'd have a professional realtor help you prepare a real-estate contract.

An alternative way to approach the partial ownership structure is to:

- Form a company in which you are the controlling shareholder;
- Assign your patent to the company;
- Grant your part-owner a profit share in the company.

That way, you control the disposition of the patent completely, and your new partner gets the same monetary return as he would have if he had part ownership in the patent.

Contract law varies from state to state. In most cases a verbal agreement constitutes a legal, binding contract. However, verbal contracts have some limits, so it's highly recommended that you do not go even one step forward with a handshake agreement. Get it in writing and have it prepared by a qualified attorney.

Very frequently, joint owners have bitter, irreparable disputes somewhere along the way. They might have started out as best friends. Keep that in mind as you contemplate assigning part of your rights. Your enthusiastic partner today could be your aggressive adversary tomorrow. It's truly very much like a marriage.

If you simply cannot go forward on your own resources but strongly want to remain part owner of the invention, partial assignment is one way to do it, but it's not a very good way. The headaches and loss of control that usually come with co-owners simply aren't worth it. It's better to sell outright and get on with your life or choose a different route, such as licensing, or partnering to acquire capital or services.

Please go to the Quiz on page 259.

You should now be able to respond correctly to quiz items 107 through 114.

Patent Licensing

Licensing is another common way an inventor can realize early gains from his inventions. Apart from fully assigning ownership, it's usually the way that requires the least investment of time and money.

Whereas an assignment is the outright sale or transfer of intellectual property, a license agreement is analogous to *leasing* legal rights to use the property for some length of time. License agreements are generally much more complicated than full assignments because, like partial assignments, they define an interactive, ongoing, possibly long-term relationship.

Normally a license agreement is between two parties: the inventor (the licensor), and some entity that wants rights to use his invention (the licensee). Licenses typically have certain covenants that both licensee and licensor must meet to keep the license in place, and these will be examined in some detail in this section.

We start with a few things that as the licensor you must be very cautious about. They are highlighted here at the outset of the licensing discussion because they have the potential to be disastrous.

You will likely have to warrant that you're actually the owner of the technology. You should, therefore, be as certain as possible that your invention's ownership is not compromised in any way, say by your employment or by a co-inventor or co-owner. It might also be clouded by previous license agreements that were not properly terminated. If you have any doubts at all, get signed releases from anyone who otherwise might later try to claim some ownership rights.

You might also be asked to warrant that the licensed invention doesn't infringe on anyone else's valid rights. That's a harder problem. It's so difficult to examine the immense amount of patented technology that you can never be absolutely certain yours doesn't infringe someone else's. An infringement challenge could happen anytime during the lifetime of the license, leaving you perpetually vulnerable. So, I recommend you do *not* indemnify the licensee against infringement upon the rights of others. The licensee should do a diligent enough search himself to be satisfied that what he is about to license doesn't infringe; then he should take all of the responsibility if an infringement challenge subsequently occurs. Of course, you should have made a thorough search yourself prior to offering the license, and you will, in any case, probably have to swear that you know of no actual or potential infringement.

Unless contractually limited, indemnifying the licensee against losses due to legitimate claims relating to the abovementioned warranties could be financially devastating. That's fairly sobering. If you do indemnify, be sure your exposure is limited to some tolerable amount.

The same is true for liability for physical harm or financial losses resulting from the use of licensed products. That responsibility should also be entirely shouldered by the licensee.

Of course, there will be some covenants the licensee will make to you, such as how and when you will be compensated for the license, performance goals he will have to meet, and so on. These will be addressed in the context of the following detailed license agreement discussion.

License agreements can take many forms and, like land leases, can be whatever the signatories agree upon. Generally they define the payment of money to the inventor for the right to use his invention. They are often exclusive, meaning that the licensor grants use rights to a single licensee, but they don't have to be. Suppose the invention can take on several embodiments that would find use in different industries. In our earlier Figure 7 example, we described a geared bottle opener. It's clear that it could be produced in one form for kitchen use and in another larger version for, let's say, unscrewing oil filters on cars. As the bottle-opener's inventor, you might license one company to produce and sell the kitchen version and another to do the same for the auto industry.

As a beginning, independent inventor, you probably don't have the resources or know-how to produce and market your invention yourself. You'd need engineers to generate the formal design drawings and manufacturing procedures, a factory to produce the product, and a sales network to distribute it.[1] So granting a license to a company that already has everything in place to get your product out is a good option to consider.

Finding a Licensee

Your first task would be to identify candidate licensees. Continuing with the geared bottle-opener example, here are some of the ways you could do that:

- One very simple first step would be to go down to your local supermarkets and gourmet shops to see what bottle openers are on their shelves. Ask the shop owners which products sell best and which draw the most complaints. Look at the products. See how they're packaged. Check out their quality. Try to judge which manufacturers you might feel comfortable being associated with.
- Go on the Internet. Look at the sites of the manufacturers whose products you've examined and look at any others you can find.
- Go to your local library and/or the Internet. Look through cooking magazines for advertisers. See if there are any professional trade magazines for the relevant products.
- See also if there are any trade shows in the field of your invention. Trade shows are a great place to see leading-edge technology.[2]

In summary, just try whatever avenues you can think of to get some feel for the manufacturers in the industry you wish to enter.

Once you've done that, pick a few candidates you'd find acceptable as licensees. Rank the list. Then begin to work out strategies for contacting each of them.

1 See the section on SWOT Analysis. Use it as a tool to help you gauge your capacity to do it yourself.

2 A useful site for trade show calendars is found at www.tsnn.com/.

Talk to them all before choosing a licensee. Start with your least favorite. You'll become better at presenting yourself and your technology through practice. So save your favorite candidate for last. It will be your best presentation.

You'll find many companies, particularly larger ones, will be reluctant to talk to you. That's especially true if your invention is not already well defined by an issued or pending patent, and that's one reason why it's not advisable to seek associates earlier in the game. In even the best of circumstances, it's very hard to get the attention of a good prospective licensee. If the candidate has to sign agreements not to disclose your proprietary data, he could be put off from agreeing to see you. Also, the candidate will worry that you could disclose information relative to development work that his company already is contemplating or has in progress. He knows he could be left open for litigation from you downstream.

The candidate will, in short, probably not be willing to accept any proprietary information from you whatsoever. In fact, he will probably ask you to sign an agreement *not* to present any confidential material. It's clearly better if you can approach the prospect with a well-defined, patented innovation that can be scrutinized with no worry of possible unpleasant repercussions to either party.

You'll also find that many companies are not interested in ideas that haven't spawned within their own group. Corporate engineers are especially hard cases when it comes to accepting outside technology. It's possibly team spirit, or maybe pride, but they often think no one can do it better than they can. Frequently they'll reject an outside idea without even considering it. It's called the "not invented here" syndrome and it's very real. They are protecting their turf.

In my experience, the best initial contact that the outside inventor can have within a prospective company is the marketing manager. After all, it's he who has the responsibility to broaden the company's markets, and in most companies it's he who guides the direction of new-product development.

So, as you're trying to establish an inroad for presenting your ideas to a candidate licensee, try to catch the attention of the marketing manager.

Failing that, try the chief financial officer. He, too, has his eye on the company's profits and does not have as much resistance as, say, an engineer, to accepting outside technology. It is no reflection on either the marketing manager or the CFO if their own development team has failed to come up with the good new idea, but it looks bad for them if they don't make money.

It's best to make your first approach in person rather than by phone or letter. Of course, that's not always possible. It usually takes at least an initial phone call to set up an appointment. Try to limit your first phone conversation to just that: making an appointment. Before you make contact, be prepared to establish your credentials. That is, be ready to briefly state what your invention is and the experience or training that makes you a credible designer of bottle openers (or whatever your innovation might be). Know what you're going to say before you call.

Before your appointment, prepare a discussion on your technology. Unless you are simply awful at such things, you should present the material yourself. No one's a better ambassador for the technology than you, its inventor. It's you who have the most enthusiasm for the ideas. You have the most insight into where the technology can lead and what adaptations are possible. It's you with whom the licensee will eventually work in bringing the product to the market. As you make your presentation, the potential licensee will be sizing up both you and your invention. You will in some sense be interviewing for a spot on his team.

If you don't capture your candidate's interest in the first minute of your presentation, you probably will have lost it forever. So, concentrate most on your opening. If at all possible, make some representative, functional prototypes of your device. When you go into the meeting, don't begin until all attendees are present and settled down waiting for you to start. Be patient. Take your own business cards and ask for theirs. I usually array their cards on the table in front of me in the order in which they're seated. It helps me keep track of who's who. Study their cards as you receive them and try to remember what their respective positions are within the company. Someone in the group will be the decision maker. Try to ascertain who that is and play to him.

Lay out your prototypes on the table, out of their reach, before you begin speaking. Now you have their attention. No one can resist being curious about a new gadget. Explain the virtues of your innovation and then pass your prototypes around as you continue speaking. Be succinct. Be truly grateful for the time they have taken to meet with you. They're busy people.

Remember, your candidates have only one goal in mind: making money with minimum hassles. So, as you organize your discussion, you should keep it simple and emphasize the following points:

- Why you are there: To explore a licensing relationship with them
- Why you have chosen them as a possible licensee
- The principal unique features of your innovation and how they work
- How these features will set your product apart from their competition
- Why many customers will choose your product over others
- What segment of the candidate's market your innovation will address

- How the candidate will make money by licensing your technology.
- When you close your presentation, try to draw out some reaction. Try to make a firm appointment to begin discussions of the license agreement itself. If you cannot make a follow-on appointment at this moment, it probably will never happen.

There's another, possibly even better, setting to get your invention in front of prospective licensees. It's the trade show, a venue that I have used myself very successfully. As mentioned earlier, it's a great place to learn about new technology, but it's also a good place to meet industry leaders. Almost all industries have trade shows at least annually. Major players in the industry have exhibition booths there. Minor players are in circulation. The booths are typically manned by the company's sales staff, but some of the company's top executives will also be at the show. It's a chance to approach them in a casual setting where they're more than normally accessible. When you go to the show, take your presentation and prototypes. Exhibitors often have private meeting rooms for such purposes. It's a pleasant, informal way to get your ideas in front of the right people.

One disadvantage of the shows is that it's hard to get all of the relevant people from the target company together at the same time. They're out making the rounds, doing their other business. So, you might have to give your presentation a few different times to get to all of them. Again, be patient. It will be worthwhile.

You might not attract a candidate licensee on your first few tries. Eventually though, one will probably be enthusiastic about your invention and want to go forward with further license agreement discussions.

Option Preliminaries

The interested candidate will probably first want an *option* on a license. That means he'll want a signed contract that allows some time, sixty days or so, to make a decision about licensing your intellectual property. The option taker will expect to pay you some agreed-upon sum for that privilege.

The option agreement may or may not include the full terms of the proposed license agreement. We'll discuss shortly what some of those terms should be. Keep in mind that if you grant the company an option to license, you're obligated to grant the license if the company picks it up. So, the option agreement should contain as many of the eventual license's important points as possible.

You can move things forward by arming yourself with a "term sheet" describing the main features you hope for in the actual license agreement. It will serve as a convenient set of talking points as you discuss the option. You should have an attorney help you with the term sheet, and a pro-forma license agreement to keep in your back pocket in anticipation of option discussions.[3] Setting out the terms of the option contract is a tricky thing to do, as the option taker will not as yet have had a chance to fully evaluate your intellectual property.

The option agreement should state that both parties will exercise "best efforts" to agree upon the final license terms. Try to give yourself a graceful exit if mutually agreeable licensing terms turn out not to be possible.

During the option period several activities will take place:

3 One reviewer of this book criticized its frequent recommendations that the reader seek professional legal help. Although it poses an expense, at certain points along the way if actions are not done just right the inventor could lose it all, and then some. So I continue to urge the inventor to do things in the safest possible way. This book is no substitute for professional legal assistance.

- The licensee's company will evaluate the manufacturability of your device.
- It will make estimates of production, packaging, and selling costs.
- It will also be trying to judge how much more development work is necessary to make the product market-ready.
- Its sales team will be evaluating the potential revenues for your product and projecting what the profits might be.
- The company's engineers will be trying to shoot holes in your design. They'll be trying to see how they can accomplish the same thing your invention does, without employing your patented technology. And, if they're cooperative, they'll be seeing how to best utilize your technology in their systems.
- The company's attorneys will be busy on two fronts: (1) scrutinizing the smallest details of your issued or pending patent, and (2) trying to craft a licensing agreement most favorable to their company. They'll be examining the details of your patent's claims, not with the notion of cheating you by avoiding them; you hope. More likely than not, they'll be looking to see how much protection your claims actually offer. They'll want to be positive that they're really licensing something that gives them a secure, proprietary position.

License Preliminaries

If the option period goes well, the next step is the license agreement itself. Let's say you want to license your new bottle opener to BotOp Industries (BOI). The licensee will probably want sole rights to manufacture and market it into at least one specified market sector. That allows BOI to spend time and money further developing both the invention and the market for it. Don't forget, BOI will have considerable early cost tooling-up to produce, advertise, and distribute the final product. All that normally represents a substantial investment that BOI isn't likely to make without exclusive rights.

BOI should have a keen desire to acquire the license. Otherwise, it's no good for either party. It's important that the BOI employees who will be directly involved with the invention's further development are also enthusiastic about it. As noted, engineers seldom share management's fascination for outside ideas and might not put much effort into developing it, possibly even torpedoing it.

As you make your various presentations and negotiate agreements, do not take on the subordinate hat-in-hand demeanor of one who's desperate to license his first invention, even though that's possibly the case. If the licensee doesn't see your invention as a potentially important part of his business, you really don't want him to take a license. If he does view it as important, you will become a valuable member of the team, deserving his respect.

In initial discussions, you should ask BOI to produce a pro-forma business plan that indicates how it intends to bring your product to market. The plan should state what human and financial resources will be dedicated to your invention's further development. It should include a timetable of events. And it should project what the expected quarterly revenues will be for the first three years. You need to keep in mind that for the first year or more there will probably be little or no revenue. It generally takes at least that long to introduce a new product to the marketplace. Nevertheless, the plan will allow you to see how serious BOI is. As we'll discuss shortly, it will give you a basis to establish certain covenants BOI must meet to retain the license.

Basic Elements of a License Agreement

A few of the most important license-agreement elements are mentioned here; however, licenses are typically very complicated, case-specific agreements. Their many aspects go well beyond this book's scope. There are many self-help books that do purport to show you how to construct a do-it-yourself license agreement.[4] I highly recommend you do not attempt it yourself. You can be sure your licensee will be represented by a trained legal professional who will be negotiating against you. As you read this section of the book, note that there are many nuances to such agreements that could cause you major grief if not written properly. Again, you should engage an experienced attorney to guide you through the process and draft your agreement. Most likely the prospective licensee will offer to have his corporate attorney draw up the agreement. Unless the licensee absolutely insists on that, though, it's better to have your own lawyer do it. That way the basic document to which negotiated changes will be made starts out favorable to you. At the very minimum, you should keep your own legal counsel intimately involved every step of the way.

A licensee typically will require certain privileges to transfer his license to another entity. That's likely to be disagreeable because you've chosen the licensee with some care and don't want an unknown new partner. But the licensee's company might be sold, for instance, in which case it would want transfer rights to the buyer. One can imagine other legitimate transfer circumstances, so some flexibility on your part is required.

In another scenario, the licensee might find it desirable to sublicense the rights to another company. In some cases that could be favorable to you. For instance, it might broaden the market coverage. As an example, suppose you

4 Be very careful using self-help books on license agreements. Due to recent U.S. patent regulation changes, all that I have seen are out of date. Check their publication dates. Major changes occurred in early 2013 with the America Invents Act. Prior publications will not reflect them.

grant an exclusive license to BOI to manufacture your patented bottle openers. If BOI sells only into the household market, it might wish to sublicense to another company that addresses the automotive market. Or it might want to sublicense to a firm that has a strong European market. Each of these would seem desirable. In other cases, however, it might be unfavorable, possibly leaving you with an undesirable partner. Your agreement should specify under what conditions the licensee can sublicense.

There is a fundamental difference between *transferring* a license agreement and *sublicensing* the agreement. A licensee who transfers rights to another entity can remove itself from the picture completely, leaving you to deal with the transferee for payment and accountability.

On the other hand, in granting a sublicense, the original licensee remains responsible to you for these activities. Be careful of that distinction in your agreement. It's potentially very important to you.

Here are some fundamental license-agreement points dealing with how and when you will be paid.

Option Fee:

The prospective licensee should pay you a non-refundable fee at the time the option agreement is signed. The fee reflects the loss of opportunity and possible revenues that you'll incur by keeping your intellectual property off the market for the option's duration. The option period should be about 30 days and not more than 90 days. Under normal circumstances, it shouldn't take the licensee any longer than that to decide whether or not to exercise the option. Keep in mind this might be just the first option you'll grant. If it isn't exercised, the next candidate will want one, and so on, sequentially draining your property's lifetime.

Don't expect to make your money on options. Unless you have evidence to the contrary, you should presume the licensee is acting in good faith and will diligently perform whatever it takes to make a timely decision. That means he's spending some time and money thoroughly evaluating your invention. That's just what you want him to do. The option price depends on how hot the invention is and how long the option lasts. For a sixty-day option on the bottle opener, something around $10,000 to $15,000 would not be extraordinary.

Royalties:

They are payments to the licensor for the right to use his intellectual property. They are usually based on some percentage of sales of the licensed technology. It's advantageous to the inventor to have that percentage based on "gross" sales, where gross sales equal the total dollar amount invoiced to customers. The licensee, conversely, will probably want to base the royalties on "net" sales, which are the gross sales minus certain deductions to be defined and agreed upon. Because there's no standard definition of net sales, using that as a royalty measure gives the licensee a chance to sweep in deductions favorable to him. Gross sales are a lot easier for you to define and scrutinize than are net sales.

There is no set formula for what the royalty percentage should be. It depends a lot on what the technology is and what your contribution to the final product will be relative to what BOI invests to develop and market it. Going back to an earlier example, suppose you brought a straightforward, simple, high-profit-margin product such as a new doll to BOI's Toy Division. BOI would need only to outsource the fabrication, and package and distribute the dolls. Little or no development is required. Your contribution would be relatively large and BOI's risked investment relatively small. In that case, your royalty would probably be higher than average.

On the other hand, let's say your invention is a new freight-train-engine crankshaft. You license BOI's Train Engine Division the rights to it. They then would have to create the production design, build and test prototype engines, develop the market, and many other things. Their investment would be huge and your royalty percentage relatively small.

Suppose your invention is truly revolutionary. Some years ago a few researchers announced, wrongly, that they had produced "energy in a teacup." That is, they claimed a significant energy source containable in a small volume and produced from the benign chemical reaction of common inexhaustible materials. If you had an invention like that, it would completely change the world's political and economic landscapes. In such a case, even though a tremendous amount of development would still be necessary, the royalty should be very high. You'd be bringing to the table a key enabling technology. Without your contribution, the breakthrough would not be possible.

The spectrum of relative contributions from licensee and licensor is clearly very broad. The best advice is to do your utmost to be realistic as to where your innovation fits within that spectrum and to be reasonable in your expectations. In most cases, the licensee has the most expenses in bringing the new product to the market and is therefore the most at risk. Don't ruin a potentially good agreement by starting out too high in what you ask. Decide the minimum return that would make you happy. Then up that by ¼ or so and negotiate within that range.

The negotiated royalty will be based on what BOI thinks it can pay and still be profitable and competitive, balanced against what you are willing to take for your rights. Royalties typically range within a few percent of gross sales, say 3 percent to 6 percent. Royalties greater than 10 percent are rare, but do occur, particularly for some software and Internet-related patents.

Royalties are often staged, meaning they might be a certain percentage on the first $1 million sales, slightly different on the second, and so on. Again, there are no real firm guidelines on the negotiated amount.

Royalties aren't always straightforward to calculate. Suppose you invented a new type of valve that is useful only on small engines that BOI's Small Engine Division and others make. The valve greatly improves the engines, and whoever gets to use your valves on its engines would have a keen competitive market edge. BOI gets the valve license and upgrades its engines. There are no actual sales of the valves, however, just sales of engines with the valves as component parts. There could also be sales of BOI-built vehicles whose engines incorporate the valves. One way to approach the royalty question in such circumstances is to assign a value-added amount to the valves themselves and then base the royalty on the number of valves BOI draws from inventory. You would have to agree beforehand how much value the valves add to the finished product. It would certainly be far more than the actual value of the valves themselves.

Royalties should be reported and paid periodically according to an agreed upon schedule.

However the royalties are calculated, you must have the right to inspect the licensee's records to ascertain that they are correct. If the licensee doesn't report or pay on time, or fails to give you timely access to the relevant records, you should have adequate remedy built into the agreement. Normally that means that if, after a reasonable period of notification, the problem hasn't been corrected you can opt out of the agreement and the licensee must adhere to all of the pre-agreed termination conditions.

Minimum Annual Royalty

At the first anniversary of your license agreement, and every anniversary thereafter, the licensee should be obligated to pay you a guaranteed, minimum, non-refundable advance against royalties for the upcoming year. That is, regardless of what the sales are for that next year, you're guaranteed at least that minimum amount. If earned royalties exceed that, you will be paid the excess. If they're less, you don't have to give anything back.

Here's one place where the licensee's pro-forma business plan comes in handy. In your initial discussions, you will have asked the licensee to prepare a business plan, including projections of annual sales over the first few years. Now you can go back to him and say, "You have projected annual sales of $1 million (as an example) averaged over the first three years. I'll allow that perhaps you've overestimated by 30 percent; so a typical year's sales could be as low as $700,000. Our agreed upon royalty is 5 percent giving me a minimum expected annual return of $35,000. Let us set the minimum annual royalty at that." If the licensee says it's far too much, he's admitting he has no faith in his business plan. That's cause for alarm.

The minimum annual royalty is a very important aspect of the agreement. It should be high enough to cause the licensee acute pain if he's not actively marketing your product. If it isn't paid, you must have the option to terminate the agreement. The minimum annual royalty, therefore, is a strong motivation for the licensee to give your product priority, and it's your way out if the licensee's not performing. Otherwise he could keep you on hold while your patent's lifetime ticks away.

Advance Payment:

A non-refundable advance payment is usually paid to the inventor upon signing the license agreement. The amount should be about equal to the minimum annual royalty. In fact, it's usually, but not always, treated as the first year's minimum annual royalty. There might be extenuating circumstances at the relationship's outset that favor a different up-front form of payment. Nevertheless, it should be something close to the minimum annual royalty. Without that, the licensee can sit inactively on your product until the first year's minimum royalty payment is due. If he doesn't pay it, by then everyone has wasted too much time and your patent has aged another year.

Technical Support:

Unless the application of your technology is completely straightforward, as in the toy's case, the buyer probably would seek your continued help to get the product out. You should encourage that. It's actually good for you, because you can continue to exert some influence on the product's further development, perhaps keeping it from getting stalled.

Your participation could take the form of an employment or consulting agreement. You can establish what professional consulting fees are in the licensee's area and use the high end as a negotiation starting point. Such information is often available from the local chamber of commerce. Otherwise, you could negotiate a fixed fee for certain specific tasks requested by the licensee. However you do it, be sure you're compensated adequately for your time and expertise. Remember, you have something unique to offer. You're the expert on your invention.

Don't forget to exercise caution in executing any non-compete or other limiting agreements required for your ongoing efforts. Your agreement not to compete has a value in and of itself, and you should be compensated adequately for that alone.

Improvement Rights:

An emerging technology typically matures through a sequence of improvements. These can be significant, sometimes eclipsing the original idea and even taking it into completely different fields. It is therefore vitally important that any agreement relating to your invention clearly specifies how improvement ownership and rights are determined.

Here are some things to consider:

- It's common, and important, that improvements whether made by you or the licensee become your property.
- If improvements are made by the licensee, they normally fall within its license, possibly with a re-negotiated royalty. If the license terminates prematurely, all improvement rights should inure to you.
- If improvements are made by you, they would normally also fall under the license, providing they are relevant to the licensee's field.

Take the invented bottle opener as an example. If the license is restricted to home use, as opposed to automotive or other markets, and the improvement relates to home use, the improvement would normally fall under the license. But, if your improvement relates to another industry, it wouldn't. Your agreement should allow for a renegotiation of royalties for significant improvements you make. Again, be careful not to give anything away for free.

Maintenance fees:

Your license agreement should state that the licensee shall be responsible for the payment of all patent maintenance fees.

When the License Terminates:

At some point, the license will end either by a breach of the agreement or by simply reaching its term. When it does, if the last-issued patent is still active, the agreement should obligate the licensee to stop production immediately and wind up licensed-item sales. Your agreement should further specify the disposition of all materials related to your invention. All drawings, molds, procedures, proprietary technical data, samples, special tooling, and so on should become your property and be promptly made available to you. The licensee should have no further use for them. That insures the licensee's production will cease and gives you some material to continue your efforts. You should be paid all royalties due you. If legally permitted, all of their agreements to keep your proprietary information confidential should remain in force perpetually.

Partnering

Instead of assigning or licensing some or all of your intellectual property, you might choose to partner with another entity or to attract partners to form your own new business. Either way, you could acquire capital to further develop your invention and perhaps get an early financial reward for your own efforts. Business partnering offers lot of options because many so different business structures are possible.

You should consider all of the common ways to partner in business before deciding what's best. For instance:

- You could partner with an ongoing small business that's already set up to manufacture products like yours. That's an alternative to licensing or selling your invention to them. In return for your invention, you would acquire an equity share in their business.

- You could seek a synergistic partner who might have technical or marketing expertise and funds to bring to the team.

- You could find a purely financial partner. In addition to funding, a venture capitalist would most likely bring some business expertise to the table. That combination could open other options for you.

- You could set up a formal entity such as a limited partnership (or corporation) and sell interest to numerous investors: family and friends, for instance. Your intellectual property could be assigned to the new entity. The resulting entity could then start a business to manufacture and market the invention, or it could simply use the funds to advance its development and more aggressively pursue any of the other options. Starting a new business to produce and market your invention is treated much more thoroughly later in the book when we arrive at Map point [25].

If you do take on partners, don't underestimate your own contributions to the partnership. Inventors frequently undervalue the importance of their technology relative to the financial or other support provided by their partners. Make sure you retain a fair share of the ownership commensurate to the value of your technology and expertise.

No matter how you partner, you'll lose some flexibility to do things your own way. Therefore, it's strongly advised that you retain at least a majority interest in the partnership. Completely losing control of one's own creations is very painful indeed, as I have learned on occasion.

The factors to seriously consider if you take on partners are pretty much the same as those outlined in the previous section on the partial assignment of patent rights. Please re-read that section if you are considering partnering.

Please go to the Quiz on page 259.

You should now be able to respond correctly to quiz items 115 through 132.

(No Model.) 2 Sheets—Sheet 1.

N. TESLA.

ELECTRIC GENERATOR.

No. 511,916. Patented Jan. 2, 1894.

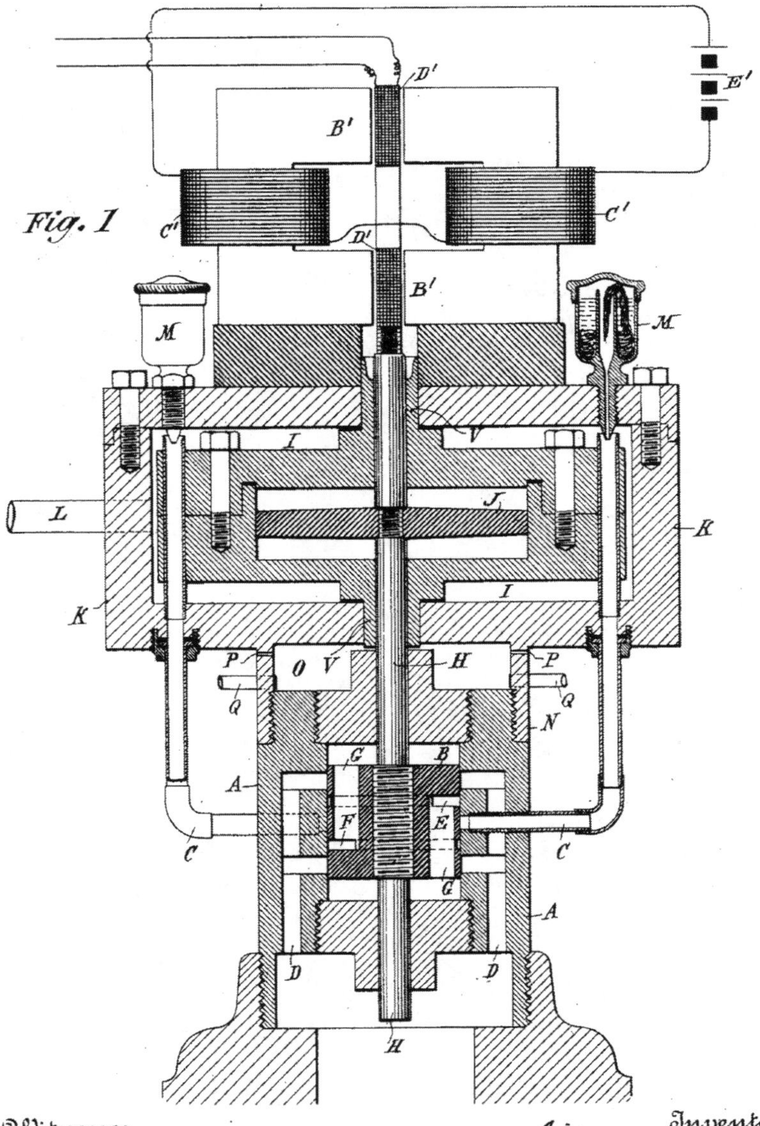

Fig. 1

Witnesses
Raphaël Netter
R. F. Gaylord

Inventor
Nikola Tesla
By his Attorneys
Duncan & Page.

14 A SHORT DIVERSION: SOME PERSONAL EXPERIENCES

We'll get back to the Map shortly and discuss the fourth way mentioned to cash in on your invention: starting your own business. Before examining that option, though, let me relate some of my own experiences. Maybe there is something to be learned from them. I've been developing intellectual property for a long time, and except for partial assignment, I've been through each of the previously mentioned ways to profit from my inventions more than once.

In the beginning, inventing was simply a satisfying avocation for me. But it drew more of my attention as time went on, and finally, about half-way through my career, it eclipsed my oceanographic research work.

Most of my patented inventions have been electrical and fiber-optic connectors and ancillary components for subsea power and communication networks. Underwater connectors work pretty much like the ones used to plug-in your home extension cords, but they can be plugged and unplugged in the ocean. Frustration with the poor products I had to work with in my early oceanographic work moved me to create better ones. As you will see, my inventions are primarily mechanical, so my approach to inventing is from that direction. But the basic steps are pretty much the same for any sort of invention.

Figure 11

My first modest success in commercializing a newly invented product happened in the 1960's. As a young oceanographer working for the U.S. Navy, my experiments were conducted from the oceanographic research platform shown in Figure 11. It was called the NEL Tower.[1] Its structure towered upward from the sixty feet deep seafloor a mile offshore of Mission Beach, California. Hundreds of electronic sensors that measured various properties of the seawater and sediment, such as temperature, salt content, and water

1 The U.S. Navy Electronic Laboratory's oceanographic research tower; its development and utilization. San Diego, Calif., U.S. Navy Electronics Laboratory]1965. Find at: https://archive.org/stream/usnavyelectronic00lafo#page/n3/mode/2up

motion, were arrayed on the tower's legs and on cables buoyed upward from the surrounding seabed. One of my jobs was to keep them all working. That involved a lot of underwater maintenance.

The use of electronics in the sea was still pretty much in its infancy then, and no good way existed for connecting or disconnecting the instruments underwater. So to replace or repair a sensor, it and all of the cable joining it back to the topside instrument laboratory had to be pulled up, refurbished or replaced, and reinstalled. There were some underwater-pluggable electrical connectors available. But they were so unreliable that they created nearly as many problems as they solved.

I was one of a handful of civilians in the 1960's qualified as a U.S. Navy diver. My diving teammates, when I wasn't diving alone or with other civilian Navy divers, were frequently ex-members of the Navy Seals, or of the Navy Underwater Demolition Teams (UDT). Much of our underwater work was done using scuba equipment or clumsy helmets; but I preferred, and often used, the lightweight "Jack Browne" system. It was designed for the U.S. Navy during World War II, and was one of the most widely used pieces of commercial diving gear then in existence. The rig consisted of a full-face mask with an umbilical line leading from the mask back to an air compressor on the surface. In my case it went either back to the research tower or to a dive boat. A person on the surface managed the umbilical line and controlled the air supply. I always dove alone when I could, something that's probably not possible with today's rules. Actually, it wasn't really permitted then, either; but most of us did it anyway. I liked the solitude of diving alone, and I didn't have to worry about looking out for a diving buddy.

In the course of five years I made nearly a thousand dives under and around that tower to keep things running. It was not the sort of diving you might dream about. Dragging heavy, barnacle-encrusted cables in murky, sometimes frigid water is grueling, dangerous work.

So as it turns out, that miserable experience guided me to the doorstep of an unsolved, and I thought solvable, problem: There was a need for electrical connectors that could be reliably and repeatedly mated and unmated underwater. I put my faith and energy into finding a way to fulfill that need.

To skip ahead with the story a little bit, it was in fact possible to do; unfortunately, not in time to help me much with my own undersea work.

In any case, my discontent led to a "fundamentally good idea." It was to isolate the point of electrical contact from the seawater by moving it into a small capsule of benign, insulating fluid.

All in all, the "fundamentally good idea" proved to be solid. Its basic elements were:

- Make the electrical connection in a non-electrically-conductive fluid of choice, not seawater
- Provide that fluid in a small sealed, penetrable chamber
- Balance the chamber's interior pressure to that of the outside seawater to avoid pressure-induced leakage.

Drawings from my first patent, U.S. 3,552,576 shown in Figure 12 illustrate simple embodiments of the idea. The invention's main concepts are seen in the Fig. 4 cutaway section of Figure 12. It shows two insulated wires terminated to contact pins, and a small chamber with penetrable walls 25a and 25b. The chamber contains a contact element 27 at its midsection, and the chamber is filled with electrically insulating fluid 16 on either side. Wall 25b has been penetrated by a contact pin with a conductive tip and insulated shaft. As the pin perforated the wall, it would have been wiped clean of any debris and water. Passing completely into the insulating fluid, it then made electrical contact with element 27. The opposed pin is shown poised to penetrate in a similar way. The chamber housing had to be replaced after every use.

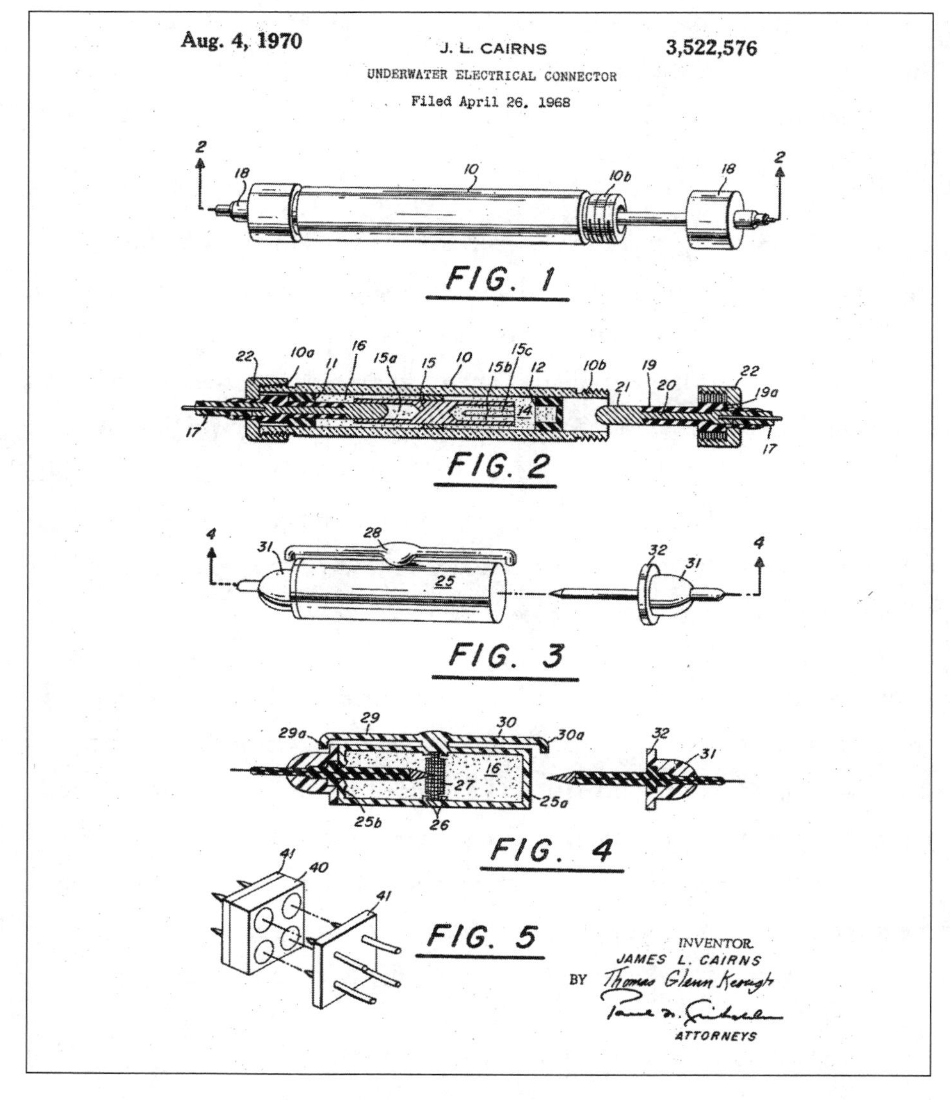

Figure 12

Fig. 1 and Fig. 2 of Figure 12 illustrate the product as it was manufactured. It connected only two wires, forming one circuit. The device was about a half-inch in diameter. It differed from Fig. 4 of the same drawing page in that there were cup-shaped rubber end caps 12 that were easily perforated by the entering pins 21. The perforated cups moved outward during mating to accommodate the fluid 14 displaced by the pins. Otherwise the cups were free to slide a bit to equalize the pressure between the insulating fluid and the outside environment. As with the Fig. 4 embodiment, after each use the chamber housing had to be replaced.

To my knowledge, the Figure 12 invention represents the first connector patent ever filed on the oil-filled and pressure-balanced concept. It contained the basic elements of what turned out to be several very successful product lines. But that was some years later. The early rendition of the design shown in Figure 12 was very weak. It worked reliably, but as you might guess, no one wanted to replace the housing after each use. Also, most submarine cables have multiple circuits, not just one. So using the connector in a practical application was very awkward. It succeeded technically but failed commercially. Understandably, customers just didn't like it.

I was working for the U.S. Navy at the time I conceived the idea. So the Navy's attorney filed the patent on my behalf, and the Navy retained a non-exclusive royalty-free license to the invention. That didn't prove to be an impediment. Once the patent was pending, I selected a few connector manufacturers as prospective licensees. I had no trouble getting audiences with them. One of them was eager to license rights to the patent, even though it was still pending. We went directly to the license agreement without any option period.

I received $1,500 as an up-front payment against the first year's royalties. That doesn't sound like much, but in 1960's dollars it was a lot to me. The licensee's company exerted a diligent effort in setting up a production line for the product, including production molds and tooling. It also made an adequate marketing effort. All in all, the licensee made a substantial investment of time and money.

Unfortunately, the product was not readily accepted in the marketplace for the reasons previously stated, and it quickly became clear that it had no future. The license agreement had a minimum annual royalty covenant the licensee had to meet. It did not meet that, so I opted to terminate the license. We did so on good terms.

I put the $1,500 away and later used it to fund the "fundamentally good idea's" next generation. It was the subject of my second patent, U.S. 3,643,207.[2] This was a more serious design. It accommodated up to four circuits, and had no disposable parts. It could be plugged and unplugged up to 100 times subsea without servicing. Unfortunately, like its predecessor, it was not a commercial success.

Those were still very early times in the development of underwater mateable connectors, and even though this second attempt failed commercially, it was an important technological step forward.

I still worked for the Navy at the time; and as before, I disclosed it to the Navy's patent attorneys so they could file a patent on my behalf. They turned me down flat this time, citing they could not see a large enough market for underwater-mateable connectors to merit the filing fees. I asked them to grant me all rights to the technology. They promptly did. I went on to file the patent application through my own private attorney.

2 To see this and any other patents mentioned, go to: www.google.com/patents and enter the patent number. The patent will be displayed.

While the patent was still pending, I approached the licensee of the previous patent from whom I had terminated the earlier license. Once again, the company took a license without first taking an option. This time I received a payment of $3,000 upon signing the license agreement. Again, there was a guaranteed minimum annual royalty to be met. In the course of the following year the licensee's company fell on hard times. It couldn't afford to get the connector into production, and by year's end it was clearly too much for them. I again terminated the license via the minimum annual royalty clause, and we parted amicably for the second and last time.

About that time my friend John Folvig joined me as a business partner. He brought the financial know-how to our efforts that I did not have. He sought other licensees for the '207 patent. In short order a major connector company took an option to license. We received $10,000 at the option's signing, which was to be the up-front license fee if the option was exercised. To put that sum in perspective, it was about equal to a mid-level engineer's annual salary at the time.

The company did take the license. After a year in which it did little to further the product, we terminated the license agreement on their failure to pay the minimum annual royalty for the upcoming second year of the contract. This was a clear case of the "not invented here" syndrome mentioned earlier. The large company's engineers simply balked at doing anything with the product. They gave up the license without any arguments and returned all relevant development material to us, as they were contractually obliged to do.

To summarize, we licensed the connector technology twice more. Each of the subsequent licenses also yielded $10,000. In 1970's dollars, the total was substantial. Each of the licenses lasted one year, and each was terminated by us on the minimum annual royalty covenant. We made some money in those years—oddly enough, on technology that never entered the marketplace.

Not all of the license terminations were amicable. Some resulted in threats of litigation, but no actual suits were filed. I should add here, that in every case the minimum annual royalty was set high enough to cause some grief to the non-performing companies. They couldn't just remain comfortably idle with the technology. If you go on to license your own inventions please keep this story in mind.

There followed a period of a few years when we did nothing with the patents. I invested my license earnings in real estate and continued oceanographic research work. Then in 1979 I got the idea for a different embodiment of the technology, this time configured as a two-circuit coaxial connector. Coaxial cables have a single central conductor surrounded by a layer of insulation. The insulation, in turn, is tightly contained in a tubular or woven outer conductor. The whole works is then covered by an outer insulating jacket. They are the sort of cables cable TV services use, only made more rugged for subsea use. For reasons well beyond the scope of this discussion, they are able to carry much higher-frequency signals than regular two-wire cables. So they're used in electrical systems that must transmit a lot of data, like television or other forms of communications.

The resulting technology was the subject of my U.S. Patent 4,373,767 filed in 1980. I thought that addressing a different market segment, underwater high-frequency signal transmission, had some merit. There was no real logic to that; it was just a hunch.

I was no longer under any employer patent agreement at that time and had no co-inventors. My ownership was therefore unencumbered. This time I did not seek licensees. Instead, John Folvig joined me again as a partner, and we set up a small business. Our goal was to manufacture and market the product ourselves. Along with a few very minor investors, who were all family and friends, we capitalized the new company with $70,000. The company was called "Challenger Marine Connectors." The coaxial connector was its only product.

We rented a five-hundred-square-foot space in a small building shared with an independent Volkswagen mechanic. The dividing walls did not go to the ceiling, so we were beneficiaries of all the noise, fumes, cursing, and loud country music of the engine workshop. Frugality was our watchword. We bought bench-top molding presses for the plastic and rubber parts. Everything we purchased was of high quality but not excessive. The first six months were spent learning how to mold, experimenting with material, writing procedures, and building the first prototype connectors. We didn't do everything well, but somehow it all came together. We began selling connectors.

To get our marketing effort started, we purchased a mailing list from a top industry magazine: *Sea Technology.*[3] We produced a flyer with a hand-drawn sketch of the product and some typed specifications. We had some nice stationery printed. A cover letter and flyer were sent to all of the mailing list's industrial contacts.

Much to our surprise, a week or two later the phone began to ring… a lot. We were getting calls from major defense contractors all over the country. One of them was Lockheed Corporation, now Lockheed Martin. They wanted to come and see us. A week or so afterward, three Lockheed executives sharply dressed in dark, three-piece suits showed up at the garage. They were visibly shocked at our miserable workplace. Nevertheless, they were keenly interested in our technology.

They asked on the spot if they could buy a full assignment of the patent and how much it would cost. We told them yes, for $1,000,000. That set them

3 www.Seatechnology.com

back a bit, but they didn't quit. In retrospect, I'm sure now that if we had persisted they would have taken it at that price. Instead, they asked for preferred-customer status in return for setting us up in a decent factory. One without fumes. We had to agree to institute formal production and testing protocols approved by them. They offered whatever help we needed to establish a proper manufacturing facility. That was, of course, irresistible for us, and we accepted. And they really helped. They taught us most of the connector manufacturing business basics. Around 20 years later, after I had long since departed from Lockheed, I received the Marine Technology Society's Lockheed Martin Award for Excellence in Engineering. It is one of the Society's top awards. Receiving it doubly pleased me because my accomplishments in the intervening years were in large part founded on what Lockheed's engineers taught me at the outset.

It turns out that at the time of the Sea Technology flyer mailing there was a major defense underwater-communications program stalled for lack of a means to connect the coaxial cables subsea. Our product solved the problem and rescued the program. After about two years Lockheed did buy our company, for just under $4 million. They paid us from the bonus they received for solving the problem.

Later in the book, when we discuss the best time to sell an invention-based business, please keep in mind this sale to Lockheed. It is a very good example of selling a dream as opposed to selling an ongoing, established business. Lockheed decided to enter the underwater connector field based upon the oil-filled and pressure-balanced connector concept protected by my early patents. They evaluated Challenge Marine on their vision of what they could do commercially with that technology, not based on any sort of historic sales data. They imagined the technical advantages the technology would bring them when placed in their hands.

Once Lockheed owned the company, they engaged a well-known national marketing firm to examine other markets for the underwater-mateable connectors. Outside of defense communications, the firm found no substantial market prospects. In particular, they found no market whatsoever for the products in the offshore-oil industry. They really missed it. It was a clear case of looking for established markets for technology that hadn't previously existed. It underlines the fact that new key, enabling technologies often must create their own market.

In the intervening years since that disappointing market survey, hundreds of thousands of underwater-mateable connectors have been put into offshore oil and gas industry use. Their current market exceeds $200 million per year.[4] They've had a major impact on that industry, allowing operations to extend into the very deep sea. That has greatly expanded the global area available for offshore oil and gas production. Previously the associated subsea systems had to be deployed and retrieved as single, large, networked units. With the new underwater-mateable capability, individual deep-water system components such as pumps, motors, valves, and sensors could be reliably added or replaced on the seafloor using remotely operated submarines. It enabled a revolutionary breakthrough in deep-water system architecture. Thinking back, changing components underwater is the same task I tried to resolve many years earlier for the underwater instruments surrounding the NEL Tower, just on a much grander scale.

After the sale of Challenger Marine I took a few years off. Then around the end of 1988 John Folvig and I teamed up once again to start a new company: Ocean Design, Inc. (ODI). This time we capitalized the new company with $100,000, a small amount of which came from the same family and friends who had joined us before. We started out with designs for new connector

4 As projected from my experience of several years ago.

products that to this day remain the high-quality standard for the industry. And they're almost all still based on the original "fundamentally good idea" disclosed in my earliest U.S. patent.

The oil-filled and pressure-balanced subsea electrical connector concept has found use in myriad subsea components including fiber-optical connectors, massive power connectors, junction boxes, cable harnesses, and remotely operated vehicle devices. You can go to **www.odi.com** to see a sampling of the wide array of products that have evolved since then.

By 2009, ODI had wholly owned subsidiaries in Aberdeen, Scotland, and Rio de Janeiro, Brazil. It had U.S. plants in Houston, Texas, and Daytona Beach, Florida. Late in 2009, it was purchased by Teledyne Technologies for more than $100 million. The company is still thriving as part of Teledyne; however, John and I are no longer part of it. All of our interest in the Company was sold with the Teledyne purchase.

As I write this, I'm as enthusiastic as ever about inventing, and hope to be able to design things for many more years. My chosen field of subsea interconnect devices is now half a century old, but I believe it's still in its infancy. There's still a lot more to be done.

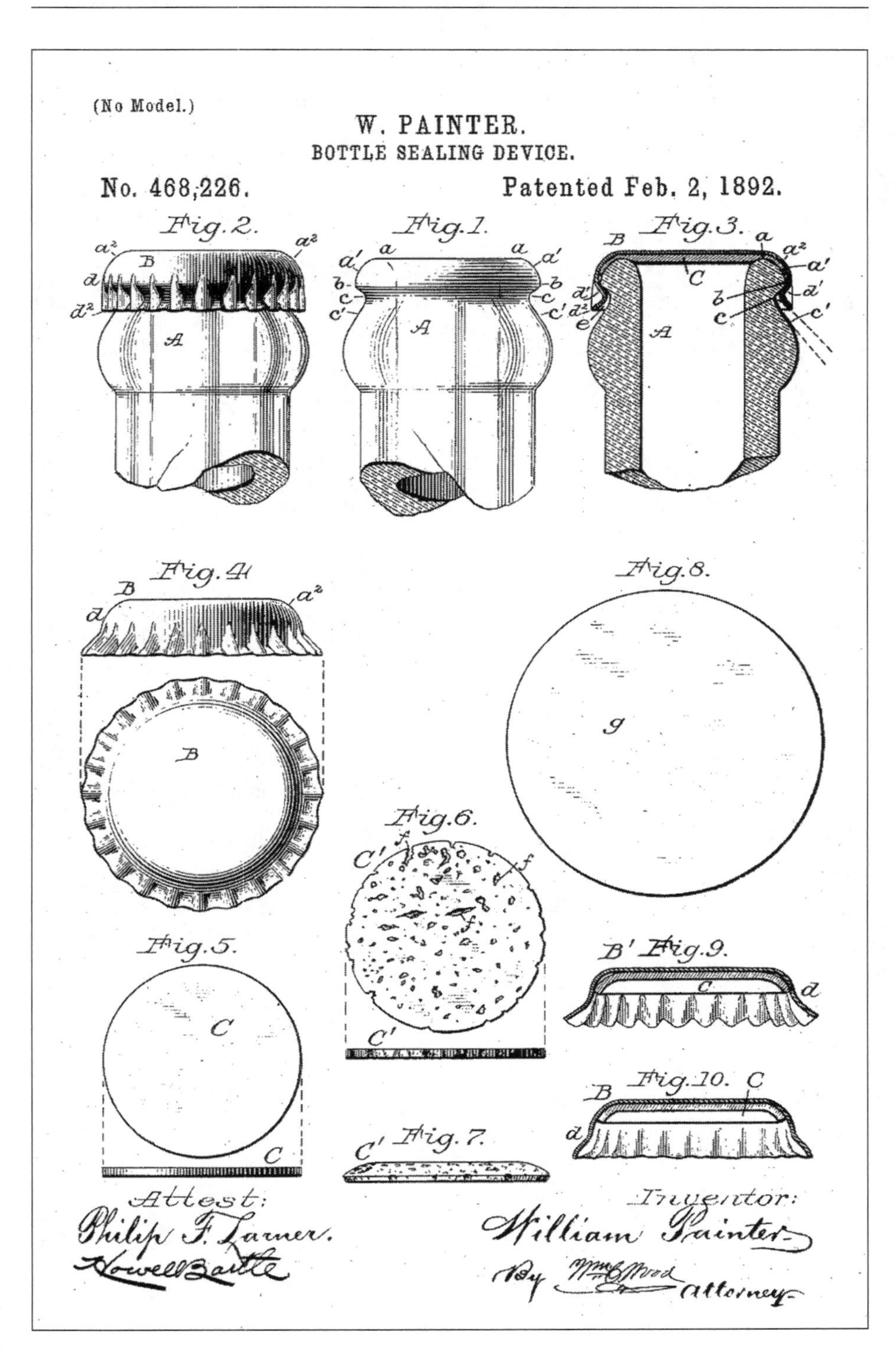
(No Model.)
W. PAINTER.
BOTTLE SEALING DEVICE.
No. 468,226.
Patented Feb. 2, 1892.
Fig. 2.
Fig. 1.
Fig. 3.
Fig. 4
Fig. 8.
Fig. 6.
Fig. 5.
Fig. 9.
Fig. 10.
Fig. 7.
Attest:
Philip F. Larner.
Howell Bartle
Inventor:
William Painter
By Wm. G. Wood Attorney

PROOF-OF-CONCEPT TESTING

You have now completed the side trip past [17], and heard a bit of my personal story. With your patent filed, and your options of assigning, licensing, and partnering in mind, you have arrived at the first really good point [18] where you can choose to cash in on your invention. Or, you can go a little further along the path to increase your technology's value with only a modest cash outlay. To advance, you must open the gate at [19] by doing some proof of concept tests.

Up until now we've concentrated on the conceptual, legal, and to some extent commercial aspects of inventing. If you have chosen to go beyond the first logical point [18] to cash in, you must now refocus on advancing the invention itself.

Way back when we started into the patent discussion, further development of the "fundamentally good idea" was put on the back burner. Since then we've explored in some detail how to establish and protect intellectual property ownership. And we've spent a lot of time looking at the various ways one can assign or license intellectual property, or form a partnership. We have been traveling along the legal and commercial segments of the pathway. We'll return to those segments later on when we discuss starting an invention-based business and even possibly selling the business. But for now, the next few portions of the journey get us back to the technical path and will go pretty quickly.

By the time you get to the proof-of-concept point, it should be clear what intellectual property you own, and you will have decided to cash in early or go a bit further to increase your invention's value. Your invention is based on certain concepts. It's time to see if they work.

Proof of concept testing is just what the name implies. It's proving the validity of the fundamental concepts on which your invention is based. It doesn't necessarily mean testing the invention itself, just the concepts that underlie its functionality. The tests are primarily to give you enough confidence to go forward with the invention's development. They need not be elaborate but should be well documented, including photos of the test apparatus. The test procedures and documented results are assets that add to the invention's value; take care of them.

Proof of concept testing is often skipped even by experienced inventors. That's a huge mistake. It is one of the most important steps in maturing your invention. If you do not do it, you will lose out on a great chance to know your invention better and as a result will probably miss opportunities for basic improvements such as design simplifications and economy. Do the tests and follow them carefully.

In cases where at least some of the basic concepts can be economically and quickly tested, it makes sense to do that testing even before filing your patent application. That way it's possible to further advance your invention's preferred embodiment before filing. It allows the resulting patent to more closely describe the final product. Expensive tests should probably be put off until after filing, though. Until then, it's still not clear that you own anything to invest in.

There follows a discussion that will give you a generic idea of how to do some simple concept testing on a mechanical device: a teapot.

A good approach is to test an invention's various concepts individually or collectively using simple mock-ups. It's better than constructing complete prototypes for a couple of reasons. One is that it permits testing fewer things at once, making it easier to pinpoint a failure's source. Another is that it's generally quicker, easier, and very much less expensive than testing the full assembly. Accurate prototypes of the complete invention will eventually have to be made. But that comes later, after the product's final configuration has been defined. Let's look at the Figure 13 whistling teapot disclosed in U.S. Utility Patent 6,386,136 to see one way that proof-of-concept testing might be done. The teapot has about the same degree of complexity as many common household inventions: not as simple as a paper clip, nor as complicated as a sewing machine. Numbers in the immediately following discussion refer to like numbers in Figure 13.

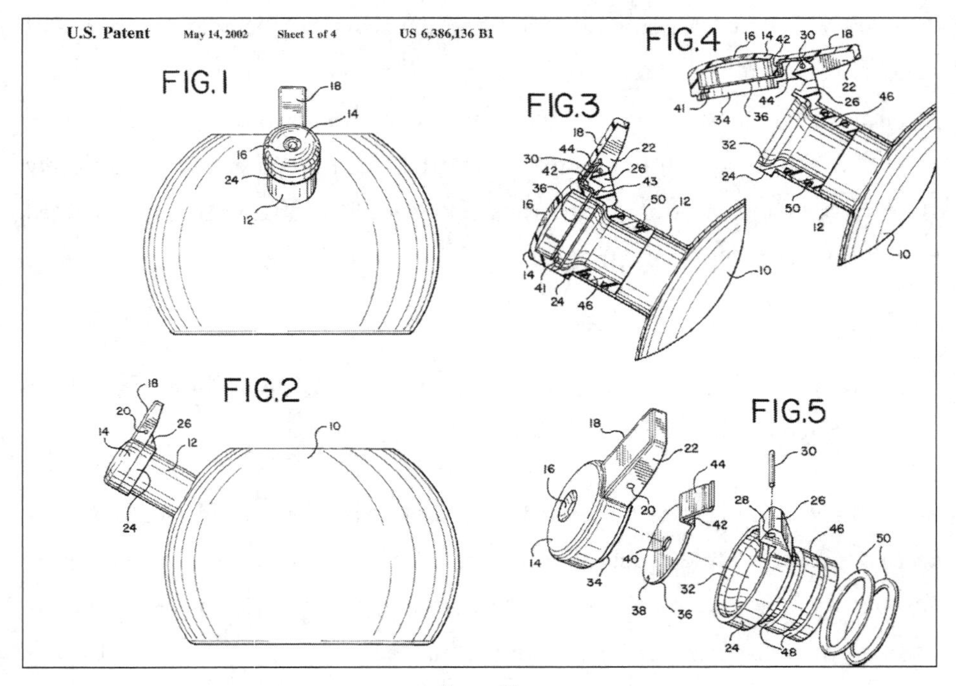

Figure 13

The teapot inventor's objectives as stated in his patent are "…to provide an inexpensive, uncomplicated tea-kettle whistle which is economical to produce, easy to assemble, has a reasonable useful life, and is still capable of providing the utilitarian and aesthetic features desired in such a construction."

To meet his objectives the teapot's whistle must perform certain functions, such as:

- When the water boils, it must whistle.
- The whistle must produce a pleasant sound.
- To function it must experience a slight pressure differential across the whistle's orifice. Therefore the Figure 13 right-hand-panel assembly, (Fig's 3, 4, and 5) must be able to sustain that pressure without leaking past o-rings 50, or blasting off into the kitchen.

- It must whistle at a relatively low pressure; otherwise the pot's lid will blow before the whistle sounds.
- Spring 44 must hold the cap in place firmly enough to sustain the slight pressure buildup inside the pot. If the cap opens before the whistle sounds, no steam will pass through the orifice, and it won't work.

The five basic concepts involved are these:

1. A whistle can be made to sound when a slight pressure differential is applied across it. The pressure gradient causes gas in the form of steam to flow through the whistle, creating a sound.
2. The whistle's sound can be made pleasant by appropriately adjusting its orifices and the cavity between them.
3. The whistle assembly shall not permit a dangerously high pressure buildup inside the pot.
4. An end cap can be held in place tightly enough by a spring to remain seated while exposed to the slight pressure differential across it.
5. The whistle assembly can be sealed into the end of a teapot's spout using o-rings.

Concepts 1 and 2 can be tested very easily and economically. The whistle's sound varies according to the shapes and sizes of orifices 16 and 40 and the nature of the space between them. To see if those parameters can be varied to produce a pleasant sound, a relatively thick washer can be sandwiched between two thin washers in a tube's end. The thick washer has a large central hole. The thinner washers each have smaller ones. Regulated air pressure can be applied to the tube's other end, creating airflow through the assembly. One can vary the orifice sizes, shapes, and spacing, and the pressure, until the right combination is found. Net cost: almost nothing. And it's fun.

Concept 3 can be tested using the same rig. This time the pressure should be increased enough to attain a very high flow rate. The whistle assembly must allow a sufficient flow rate to allow steam to be dissipated at a rate equal to or greater than its generation rate. That's all very hard to determine analytically. But the fact is, with reasonably sized orifices, testing will show that the whistle's flow rate can be made to greatly exceed the steam's generation rate. So you can do the experiment and be comfortable that the assembly can be designed to perform safely. And there's a safety backup in Concept 4.

Concept 4 is self-evident. Only the required spring force remains to be determined. However, it should be pointed out that the spring-loaded end cap serves as a backup safety valve. In case the whistle's assembly cannot handle the generated steam's flow, possibly due to blockage, the end cap will unseat to relieve the teapot's pressure before it blows up.

Concept 5 is proven a priori because the functioning of o-ring seals is well known and documented. If the seals also serve to keep the assembly fixed to the spout by way of a tight interference fit, that concept can easily be tested by making a solid mock-up of the whistle assembly's base 5 and another of the spout's bore 12. Base 5 can be fitted together with o-rings 50 into one end of bore 12, and controlled pressure can be applied to the bore's other end. One should stand well away from this experiment, as the pressurized assembly constitutes a small bomb. Such tests should be done at both the highest and lowest temperatures the assembly is expected to see in actual use. Immersing the pressurized test assembly in containers of boiling water and then ice water should just about do the trick for concept testing, although something different might be used later for the qualification tests.

The described tests are not at all rigorous, but they show that the design's basic premises could be made to work, at least to the degree tested. The assembly's end cap 24 and base 46 are intended to ultimately be made from molded plastic. Injection molds are very costly to make and difficult to modify. So simple tests such as those described could help the teapot's inventor refine his design before committing to expensive tooling.

The real point here is not to analyze just how these tests could be done, but to get a flavor of how modestly one can proceed to wring out an invention in its early stages. Working independently, it's possible to very quickly do simple experiments such as that just described. In a professional engineering company, it would probably not be so easy. Such companies have tight quality systems requiring everything to be done formally and according to certain protocols. That's not a flaw; they must have such systems. But the independent inventor's freedom gives him a slight edge.

A different approach commonly taken even by independent inventors is to create formal drawings of the invention's proposed best embodiment, have someone else make it, and then have an independent laboratory test it. That's analogous to having someone make a complete teapot, fill it with water, set it on a hot stove, and then tell you whether or not it gives a good whistle. That approach is usually a mistake. Here's why.

First of all, turning over the testing to someone else deprives you of a chance to get valuable insight into your invention's performance. You should at least participate in the tests. Getting a written test report is not nearly as insightful as being there.

Second, once those parts are manufactured you're locked into them. Sure, you could modify them, but you'd be reluctant to throw them all in the trash and start over. And, what about the molds. They'd be discarded very reluctantly.

Finally, if you were handcrafting the parts yourself from essentially free material, the test articles wouldn't be as slick, but you could scrap them without hesitation and move on to a new configuration. It's a part of the creative process you'd lose if someone else made them for you.

Even if you cannot do everything yourself, do as much as you can. It's a wonderful learning experience. If at all possible, don't miss it. As you craft and test the invention's mock-ups, your mind will be filled with ideas of how to approach the actual manufacture of the finished product. You'll be living within your design, sometimes struggling with it, but not detached from it. That's vitally important. There's no substitute for hands-on experience. There will come a time when you'll lose that intimate contact. Don't let it happen too soon.

Almost every product is eventually handed off from the designer to the manufacturer. Even if it's within the same company, design and manufacturing are typically separate groups. Once the handoff takes place, the designer loses close touch with his invention, pretty much ending the creative process. Hence the designer should be very reluctant to turn over his product until he's completely satisfied that it's ready to stand on its own. (This is also why I urge a licensor to keep involved with the licensee's company in his invention's further development. The licensee's engineers will not have his in-depth understanding of how the invention can be modified and improved.)

It's common that corporate engineers, particularly more senior ones, are reluctant to get their hands dirty. That's a mistake. They accurately generate excellent, formal, product-design packages, hand them off to manufacturing, and go on to something else. They might attend a meeting some months later to discuss why the product doesn't work well or why it's so hard to manufacture. Other than that, they are out of touch with the design. The truth is they have possibly never really been totally immersed in it. They are missing an important part of the creative process.

Whether you do the testing yourself or have it done, bear in mind there are always two things being tested. One is the device under test, and the other is the test apparatus itself. A faulty test setup will often disappointingly fail a perfectly good device. In the case of failure, one should always first suspect the test protocol and apparatus and lastly the device being tested. Unfortunately, as you will learn, customers and other independent witnesses to the tests immediately prioritize their suspicions in the reverse order. If the proof-of-concept or later qualification testing does not go well, be sure the test setup is sound. If it is, next scrutinize the test itself. Does it realistically test the concept? Or is it too harsh? Does it put under stress features of the device beyond those related to the concept? Is there something in the test protocol or in the device itself that can be modified to allow its passing?

Don't give up until every avenue has been explored. As this goes along, you'll be learning more about your invention. It's a lot better to deal with a design weakness now rather than later, and keeping your own hands in the works often lets you do that. Be careful to keep track of even the smallest changes made to the test articles or protocols during testing. Document any modifications as soon as you make them. Unless you have no choice, change only one thing at a time before retesting. That way, you know which one is the culprit. Do not wait until you have tweaked three or four things to write the changes down. The importance of keeping a carefully recorded test log cannot be overstressed. Take lots of photographs of the test setups and tested articles. When testing is complete, assemble your documentation and results in an orderly report. The report is valuable both for you and for the eventual marketing of your product.

When your invention has finally passed its proof-of-concept tests, you have opened the gate at Map point [19], and simultaneously will have arrived at the second reasonable funding opportunity [20]. Your options are those discussed earlier at the first such point. But now you have more to offer. If you wish to cash in here, and if it's possible to do so, you should build some representative prototypes of your invention. They'll be useful demonstration aids in attracting funds. A product that can be held in one's hand is a far better sales tool than photographs or computer solid models.

If, however, you want to pass this cashing-in point and go further down the path on your own, it's better to hold off on the prototypes until you've defined your invention's actual production embodiment. Product definition is the topic of our next section. Once completed, gate [21] will be open.

PRODUCT DEFINITION

Product definition is the task of choosing the initial production version of your invention. That means not only determining the marketplace into which the product will be sold but also finalizing the details of how it's made and packaged. Product definition starts out largely as a marketing activity and then quickly involves some engineering effort.

The marketing task is to find the product's most immediate and profitable markets. That, in turn, requires evaluating a number of market-specific factors including competition, entry barriers, financial risks vs. rewards, profit margins, revenue projections, and so on. It also entails weighing the start-up and manufacturing costs of the product as configured for each of the candidate markets. Fundamentally, it's guessing where to place your first bet in order to have the best chance of winning.

For the moment, suppose once again that you've invented the geared-strap bottle opener of Figure 7 for which we've done a mock patent search. As noted earlier, it's not a very good invention; this is just an example. You recognize the invention could take on several different forms for different uses. It could be used as a household bottle opener, or a larger version could be used as a household jar opener. As mentioned earlier, a different configuration would work to remove oil filters from cars. Still another embodiment could serve as a multipurpose strap wrench. For some of these applications the basic structure could be made from molded plastic. For others metal would be more appropriate. It would be a different product for each of these possible uses.

The up-front investment would be quite different for each of them. A stamped, formed, metal-handled oil-filter wrench, for instance, might be much cheaper to tool than a molded-plastic bottle opener. Molds are very expensive. So you could possibly get a metal-handled product on the market with less up-front cost. On the other hand, plastic molded parts are often much cheaper to mass-produce than metal ones, enhancing your downstream profit margin.

How do you decide which product configuration would optimize the return on your investment? All you can do in most circumstances is to make your best-informed guess. There follow some suggestions to help you make that guess.

Market research will certainly improve your chances of guessing correctly, so that's where to start. You can approach market research in a variety of ways. One of them is to hire a professional market-research company to do it for you. My experience with that has not been good. You'll recall the earlier discussion of a well-respected research firm finding no offshore-oil-industry market for underwater connectors. So while professional research firms might be useful, they cannot always be counted on to get it right. And they're very expensive. So once again, I recommend you do it yourself.

You need to find out what's already out there in each of your potential market spaces. You have to know what consumer needs those existing products address and, more importantly, which ones they do not. If you can find a niche in one of the markets that's not being serviced by an existing product, look hard to see if yours can fill the need.

Search for a market space with easy entry and potentially quick rewards. Look for one in which you will be protected from heavy-duty competition, at least until you're out of the starting blocks. Finding one in which you can quickly establish name recognition is a plus.

If your invention fulfills a common industrial need, your market research could begin on the Internet with the Thomas Register of products and services.[1] That will give you a good starting reference for your industry search. More detailed information on competitive products can be found using Google or a similar data retrieval service. By looking across all industries in which your invention might find applications, and seeing what technically similar products are used, how they're used, how they're made, and so on you will gain a good top-level level industry view that will not only help you define your product, it will also serve you later in your marketing effort.

No matter what your product, the Internet search will keep you busy for some time and will yield a lot of useful data. Here again, you might turn up potential applications for your invention you hadn't even considered.

Trade magazines provide another way to gain product intelligence. Scan those relevant to your potential markets to see what's being offered in your product area. You can also get information on industry trade shows from them. Go to some relevant trade shows if you can. It's a great way to get your hands

1 You can access the Thomas Registry at http://www.thomasnet.com/index.html.

on competitors' products. Trade show exhibitors, particularly salesmen, are always eager to show their latest innovations and explain how they work. Often they'll freely discuss their sales, best outlets, problems, and other useful information they wouldn't reveal in a different venue. You might also be able to find a distributor for your product at a trade show, if that's in your plan.

As you begin to home in on the production version of your invention that you want to launch first, go back to the USPTO's website and run another DIY search for competitive technology. You will possibly learn more about existing art that could alter features of your first product.

Gather all the data you possibly can. Then weigh that against the risks and other factors involved in each of the directions you might decide to go. From that, and your estimates of the manufacturing costs for the product's various embodiments, carefully make your selection of the product's first customer base. Remember, the worst decision is to make no decision at all. Don't let yourself get stuck here. Do your homework; then take your best shot.

As you proceed through your market research, and while the information you've gathered is fresh in your mind, continue fleshing out the sales and marketing aspects of your business plan.

Once you've targeted the market space you want to enter the next step is to tailor your product for that market's easiest user acceptance. That means further engineering to fine-tune the design's functionality and to enhance its manufacturability for that use. Engineers frequently congratulate themselves when they've achieved a design that meets its functional requirements, and they stop there. Wrong. Attorneys say, "The devil is in the details." It's as true for designs as it is for legal documents. Engineers forget that sometimes. The smallest details, if overlooked, can cause enormous grief later.

Time spent refining a design prior to its finalization is as important as any you can spend. For one thing, quality control begins there. Every single aspect of the product should be scrutinized. Is it necessary? Can it be simplified? Can you cut down assembly time and mistakes? Simple things here can make a notable difference.

Here's an example of what I mean. If your product has 10 machine screws (or any other multiple components) for instance, make them identical if possible: same material, same size. And, make them standard off-the-shelf components. That cuts down on inventory, simplifies making up kits for assembly, and reduces the chance of mounting a screw in the wrong place. If those screws go into tapped holes, machining time is reduced by making them all the same, and the number of required machine tools is reduced. Standard tooling & components are much more readily available and cost less than non-standard ones, resulting in both time and cost savings. Standard tooling can be used for other operations. You must design with these sorts of refinements in mind to get the best result.

Even though the practicality of this seems obvious, it is very often ignored. Take your time and do a thorough job. When you have finished your final design, have it reviewed by one or more experts in your field to see if you have missed anything.

When you've refined your design as much as you can, it's time to make some representative functional prototypes. They'll serve you in two ways: one is for qualification testing; the other is to show off your product. One good way to rapidly make detailed prototype components is to use 3-D printing.

With your product defined, Map gate [21] opens, and you've reached the third potential funding point [22]. It's now reasonable to cash in on your work through any of the previously outlined methods or to go still further down the road. The prototypes will allow you to present your invention in its actual functioning form in a way that shows it best. You now have quite a bit to offer prospective investors or acquirers. You have a pending or issued patent, proof-of-concept test results, some market research defining the product and its target industry, and functioning prototypes. The next milestone is product qualification testing. When completed, it will open the Map gate at [25]. You haven't had to spend much money yet, except perhaps for your patent. Unfortunately, to stay in the game beyond this point the stakes get considerably higher. If you're on a very limited budget, this is probably the best time to get out, but it's really starting to get interesting.

QUALIFICATION TESTING

There's a fundamental difference between *proof-of-concept* and *qualification* testing. The former demonstrates the soundness of the invention's various concepts. The latter shows that the product, in the form in which it will be sold, meets all of its performance requirements. When you develop something, you know what it's supposed to do, and you will probably advertise that, in fact, it does that. Before releasing it to the marketplace, you should determine that it actually does what's intended. That determination is made through qualification testing.

Qualification testing is possibly the most expensive, time-consuming step in your product's development. It is so product and industry specific that it's vastly different for every product type. You can imagine the diversity in requirements for a new bicycle wheel versus one for a commercial airliner, or for toilet-valve performance requirements compared to those for a new artificial heart valve. Their qualification test protocols will bear little resemblance to each other; however, there will be some commonality in the steps that are performed.

For instance, nearly all qualification-testing programs have certain destructive elements. They stress a product beyond its operational limits to determine the range of extreme conditions it can withstand. Depending on the product, some typical test parameters might be: high and low temperature, mechanical shock, impact resistance, vibration, high and low pressure, corrosion resistance, solar ray resistance, water immersion, number of operational cycles, and so on. Many tests are taken to the point where the product fails. Those high-stress failures reveal the product's weakest aspects. In turn, the failure data highlight ways to improve the product. Because of that, the final design modifications in a product's development are usually dictated by its qualification test results.

Depending on the invention, the qualification test criteria may be just those you choose to do yourself simply to be comfortable that your invention works the way it's supposed to. Or they may be much more complicated and dictated by government or industry standards. You are already familiar with some of these more involved tests, although you might not realize it.

You have, for example, seen advertisements in which new cars with dummy passengers are crashed into walls. That's done to prove the cars' impact resistance and passenger safety. It's not necessarily done because the automobile manufacturer wants to satisfy his curiosity about how much of a wallop his cars can take. It's done to comply with Federal Motor Vehicle Safety Standards.[1] Such tests are obviously destructive and costly.

1 Part of the Department of Transportation's National Health, Transportation and Safety Administration. Go to www.dot.gov for more details.

Suppose, as another example, that your invention were a new pharmaceutical. Before a new drug is released for public use, it has to have FDA approval.[2] And to get that, sometimes years of well-controlled and well-documented tests must be performed. That's not done because the drug companies want to reassure themselves of their products' safety or efficacy, although to some extent they would certainly do that anyway. It's done because there are regulations that require it.

Many common industrial-use products also have obligatory OSHA requirements.[3] For example, if you invented a new style wooden ladder, not only would it have to demonstrably support certain loads, but it would also have to comply with other OSHA criteria for rung spacing, sharp edges, and surface finish—no splinters.

Clearly the testing will be dictated by the product and its use. Whatever the test are, they should be adequate to demonstrate to a high degree of certainty that your product meets or exceeds its advertised specifications. Prior to finalizing your design, you should have determined what obligatory testing your product would have to undergo, and anticipated them in the design.

Even though there may be no existing regulations for your own invention, adequate testing should still be done. If you don't do it, you run the risk of returns, dissatisfied customers, and possible warranty liability. All testing, no matter how simple or complex should be witnessed by an independent, unbiased party and well documented.

Here again, just as for proof-of-concept testing, you should make every effort to stay intimately involved with the testing. Many of these tests can be very

2 The Food and Drug Administration's regulations may be found at www.fda.gov.

3 OSHA is the U.S. Occupational Safety and Health Administration. Its regulations are found at www.osha.gov.

costly, and witnessed by independent observers who will not remember that a test failed because the test set-up was wrong. They will simply remember that it failed, and that's the message they will take away with them. These tests can be very interactive, with many on the spot modifications. Test set-up faults arise: a poorly connected tests lead, high-humidity, a malfunctioning sensor, contamination, etc. When they do, they need to be recognized by knowing eyes, corrected, and documented. No one is a better faultfinder than the inventor; he needs to be there.

Adequate qualification testing will serve to support you to some extent against product liability claims in case your device harms someone's health, property, or business. Proper tests will show that you've done all that's practical to ascertain your product's safety and reliability. Notwithstanding the testing, product liability insurance is highly desirable.

Once qualification testing is complete, documented, signed off by independent observers and your product has been modified to incorporate any changes dictated by the test results, the gate at [23] opens for you. With that milestone achieved, your product is ready to be manufactured. Once again your invention's value is increased dramatically. You have arrived at cashing-in point [24] on our invention development Map, and have all of the previously described options for reaping the rewards of your invention. You are seductively close to the biggest challenge of them all, though: starting your own business to produce and sell your product. Right now, you should seriously consider taking your rewards and running. It's much safer than going forward. But if you have the funds, stamina, and talent to get your own business going, you can make the life-changing decision to do it. It is by far the most challenging and exciting part of the trip. But it is not for the faint of heart. The last gate [25] will open at your command. You just have to make the choice to pass through it.

STARTING YOUR OWN BUSINESS

This is the last cashing-in option to be covered. It is much more complex than assigning, licensing, or even partnering, and its discussion logically comes at this late point of our journey. Even the best inventions typically take years to generate substantial revenues from the sale of products. So if you're a beginning inventor, it's probably best to start with licensing or one of the other options. But, both the biggest money and the biggest challenges are just beyond Map gate [25], and at some point in your inventing career the time might be right to start your own company to manufacture and sell your product.

Succeeding with a start-up business of any sort is difficult, and many fail in short order. That's despite the fact that, like a pizza place or dry cleaners, they usually already have a product or service ready to sell. The hurdle is higher for an invention-based company that begins with nothing more tangible than an idea or two.

An invention-based start-up business guarantees you at least a few years of unrelenting pressure. You're committed to simultaneously finalizing your invention's development, setting up the physical plant, sales, manufacturing, distribution, financial, and all other aspects of your business.

It would be hard to overestimate the extraordinary physical and mental stamina needed to keep going day after day, month after month, without a break. But that's what it takes to succeed. If you start your own invention-based business, you'll need your family's full support. It won't be easy for them. For a while your new company will get nearly all your attention. The common maxim that family life should be balanced with work activities will be temporarily set aside. You won't have time to skip a beat. There will be no dull moments and no spare time. You'll be tossed about by frequent sporadic bouts of fear, exhaustion, disappointment, and euphoria. You'll be fully engaged.

But you'll love it. Consider the upside. Your potential rewards are unlimited. You'll be your own boss. You'll be challenged every hour of every day. You can create to your full potential. And your future's entirely in your own hands. It's a superb feeling.

If you take on the challenge, don't overestimate your own business talents. It takes solid business skills to succeed. Without them, all your hard work and creativity could be wasted. Line up someone who can help if you're weak in any required business disciplines. When you start, leave as little to chance as you possibly can. Make sure you have all the needed talent available.

Unless you have sufficient funds to support yourself, and to develop both the business and your technology for a period of several years, you'll likely get in trouble. Be financially prepared. All new businesses require financing in one form or another, so you'll need to build banking relationships. Start to build them before you actually need them. Establish your company's routine accounts with a financial organization that can eventually help you with broader needs. Get to know the managers. Tell them early on about your future goals. Then when you're ready to call on them for help they won't be surprised, and you won't be a stranger. They will want to see your business plan when you seek financial help for your company. They will also want personal financial statements from you and your company's other principals.

While it's true that many start-ups fail, the chances of success are greatly increased by careful planning and execution. One of the first things to be done, as mentioned earlier, is to write a business plan. It's a mistake to put it off thinking it will be easier later. Don't do that. If you don't have a plan, you're off to a pessimistic start. As noted previously, it's like beginning a trip into dangerous, unknown territory without a map. When you start your business, you have to know what you're going to do. And if you know it, you should be able to write it down. If you don't know it, you're in trouble. Go to your local Small Business Development Center for help with your business plan, even if you don't think you need it.

The plan should completely define your business including your programs for marketing, management, production, and financing. It should realistically portray every phase of your business. Set down what you intend to do, how you intend to do it, what resources you will need, how you will pay for things, and when and in what order you will do them. It should accurately reflect your goals, milestones, and potential hazards. It will allow you to demonstrate to potential investors and others that you have addressed all anticipated problems as a prelude to launching your business. As the business grows, your plan will need to be revised frequently to grow with it. Do not hesitate to modify your plan; sometimes you might have to change it from day to day, but do it thoughtfully.

To formulate and periodically update your plan, it's a good idea to systematically analyze your business. One common way to do that is through a "SWOT" analysis. A brief example of the analysis is given here simply as an introduction. There's a lot of instructional information about it available online, and quite possibly you are already familiar with it.[1]

To start a SWOT analysis, make up a four-column chart as shown in Figure 14. The chart has some example entries just to give you an idea of how to get started. On your own chart, which will probably be much longer than the example, list your company's principal **S**trengths, **W**eaknesses, **O**pportunities, and **T**hreats. Try to be as realistic and comprehensive as you can. Keep in mind the list is very subjective, so the results will be equally subjective. But it's still very useful. If there are several principals in your company, ask each to make a list privately. Then compare them. It is interesting to see if all the company's principals see the same strengths, weaknesses, threats and opportunities. An open discussion of their differences should lead to some lively, constructive debates.

The analysis is intended to help you keep focused on the main issues that affect your company. SWOT analyses are particularly useful in guiding an emerging company's activities.

Strengths and weaknesses are generally characteristics of the company itself, viewed at a certain moment in time. They are things you might be able to positively change. Opportunities and threats are factors that often exist outside of the company. Opportunities can at times be taken advantage of, and threats are sometimes avoidable. Identifying your company's strengths, weaknesses, opportunities and threats by way of the SWOT chart gives you the

1 For more discussion and some examples of SWOT analyses, go to www.businessballs.com/swotanalysisfreetemplate.htm.

opportunity to do something about them. The SWOT analysis should be an ongoing team effort, with its entries updated frequently.

STRENGTHS	WEAKNESSES	OPPORTUNITIES	THREATS
You have found an ideal location for your company	You still need to fill some critical positions in your company	There's a big contract on the horizon you might qualify for	There are still some required tests you haven't finished
Good prospects for strong patent claims	Only one year's funding lined up	Market shifts should increase demands for your products	An overbearing competitor is present in your market area
Your products will have easy customer acceptance	Company's principals not in agreement on many key issues	Main competitor is having trouble filling orders	Your operations could violate some environmental impact rules
Company's principals have strong technical skills	Your sales forecasts are based on sketchy data	Major potential customer is looking at your products	Potential customer also looking at your competitor's products

Figure 14

There is one difficult-to-assess threat will probably not appear on your SWOT chart: It's having too much business too fast. That can be as devastating as too little business. It's very difficult to quickly ramp up human and other resources while simultaneously maintaining quality.

Careful, steady team building is important in an emerging company. That's hard when sales grow too rapidly. It takes time to train new employees. When they are being added rapidly your company could wind up with a large number of poorly trained help and concomitant quality problems.

It's also a challenge to manage cash flow in a fast-growth environment. Cash received from invoiced sales might lag by several months or more the cash required to fill large new orders. The timing gap can exhaust your credit line and delay payment to your vendors. It doesn't take much for it to get out of hand.

Explosive sales are insidious; they're not quickly perceived as a threat because they're so enticing. After all, they're just what you want. Keep in mind, though, that they can knock your company off balance and quickly spin it out of control. If you see your sales are more than you can handle, raise your prices and they'll slow down. Otherwise, you could lose it all.

Business Location

It's important to carefully choose the community in which to begin your business. As you start, one by one you'll add employees. You'll invest significant time training each of them. They'll become one of your company's greatest assets. If you choose a community that in one way or another becomes unsuitable for your evolving business, you might have to relocate a significant distance away. Then you'll lose many of your employees and traumatize those who move with you. That's a heavy blow for a small company to withstand.

As you consider various locations, begin by asking the following questions:

- Does the local talent pool offer the professional and trade labor forces that you'll eventually need?
- Are there good schools? Universities? They're good resources and will help attract and keep employees.
- Does the location offer adequate means to get your products to their intended marketplaces?

- Are there local vendors for the services and outsourced manufacturing that you'll need?
- Will raw materials be easily accessible?
- Is it near a transportation hub so that out-of-town customers and vendors can easily get to you?
- Is it a community in which you and your employees will enthusiastically want to live?
- Will local labor rates allow you to be competitive?
- What is the trade union situation in the area? How will that affect your business?
- Are there any local restrictions that would impede the growth of your type of business?

Go to the local chamber of commerce in your targeted community. They're a great resource. You'll find them very helpful in providing all sorts of useful regional information such as wage scales, population age and education distributions, local synergistic or competitive businesses, schools, transportation, and so on. They can also help lead you through the permitting and business-licensing processes. Ask them if there's a local manufacturing or trade association. Such organizations provide a good way to form synergistic networking ties to other local businesses and services.

Find out if the local government is favorable to businesses of your type. If they're not, they can be a perpetual thorn in your side. But if they are, they'll welcome you and possibly even help you with some tax breaks and fast-track permitting as the business grows. Don't be timid about asking for their help. You'll be creating jobs and revenue in their community.

Also, as you search for your business location it's a good time to return to the local Small Business Development Center.[2] They are well tuned in to the local business community, and can help you decide if the area you're looking at is a good fit for your business. You will find them helpful in many ways.

Once you've selected a favorable community in which to locate, you must choose a particular business site within the area. Buying your business location is probably not advisable; you will have enough investments to make just setting up your business without investing also in real estate. Begin by leasing a site that will suit your needs for about four years. It should be in a place that will allow you to move up to a bigger site later without leaving the community. To get started you will probably not need anything beyond the basics; so don't go for the high-rent district. Something presentable but not elegant will do. If possible it should be very close to your home, as you will probably be going there often and at all hours of the day and night.

Growing Your Business

In October 2011, President Obama directed the formation of "BusinessUSA," one facet of which is an online web information service found at:
www.BusinessUSA.gov.

The extremely useful website is dedicated to the dissemination of core information regarding the Federal Government's programs and services for small businesses. Anyone with a new or growing business should not fail to visit it. It is a tremendous asset both as a learning tool and as an ongoing reference. Some of the many topics covered are:

- Creating a business plan

2 Refer back to relevant section of book.

- Choosing the structure for your business
- Obtaining required business licenses
- Financing your business
- Hiring and retaining employees
- Managing your business
- Operating your business
- Expanding your business

Taken as a whole, the website serves as a comprehensive, very instructive, course which guides the reader through every aspect of starting and growing a business. You could easily spend several worthwhile days working through the various topics the site offers. I view it as a vital online addition to this book. When you first go there study the topics that are relevant to your immediate needs; then, as your activity matures go back to it for additional instruction and references.

In addition to generic treatment of common topics, the website gives specific guidance focused on particular types of businesses, such as invention-based ones. As an example, clicking on the "Start a Business" icon appearing on that site launches a "Wizard" that asks you to define your business activity, and guides you to various sources of assistance, some within your local zip code. It will also lead you to suggested reading relevant to your activity.

For instance, when describing your activity to the "Wizard," if you click on the box titled "Did You Invent Something," you will be referred to an article titled "20 Questions Before Starting," and another called "Is Entrepreneurship for You." They are intended to stimulate you to think about a number of important business elements before you proceed.

Another useful route to follow on the website Wizard can be found by clicking the topic "Hiring Employees." Once again, you will be lead through all of the important aspects of how to add employees to your new company. The Wizard guides you to resources for recruiting, hiring, training, and retaining employees. Some time spent studying the Wizard's topics on adding new employees could save you many hours of work, and a lot of money. Here are a couple of examples.

Probably your newly acquired employees will not come in the door with all the skills needed to perform the tasks you hired them for. They will have to be trained. Spending time to train employees reinforces their feeling of importance to you and your business. It adds to their self-respect, and their loyalty to the company. It also makes it easier to recruit when interviewees know they will learn new skills on the job. That will excite most new-hires. As you bring them on board you quite possibly can get financial help for that training.

The website will acquaint you with federal on-the-job training programs managed locally through American Job Centers, and funded by the Workforce Investment Act. Your local American Job Center (AJC) can help you both recruit and train skilled workers... and get reimbursed for your efforts. You can receive up to 50% of the costs, including salaries, to provide on-the-job training for individuals you hire through the public workforce system. The AJC on-the-job training specialists will help you find the workers you need, and will also help you train them to meet your requirements. To find out how you can participate in these on-the-job training programs[3] go to: **www.careeronestop.org/businesscenter/trainandretain/fundingemployeetraining/on-the-job-training.aspx#sthash.fatEDLh0.dpuf.**

3 The website is recommended reading by the "Hiring Employees" module within the www.BusinessUSA.gov website. This is a direct link to it for your convenience.

When you're on the above website, near the bottom of the page, click on "Find an American Job Center." That will lead to the nearest location of the nearly 3000 American Job Centers nationwide where you can begin the process. Their experts are very helpful, and the programs are great. You will get to know the staff there well. It's almost like having a free Human Resources Department for your company.

Before hiring new employees, be sure to check their credentials. People can be very dishonest in portraying their background and training. I had one applicant claim he graduated with a degree in mechanical engineering from a university in Massachusetts. I hired him without checking his facts, and soon found he knew little or nothing about engineering. When I did check, it turned out the university he named had closed its doors many years earlier, long before he supposedly graduated. When I questioned him on his outright lie, he said he thought everyone fabricated such facts on their resumes. Of course, I fired him on the spot. As he was turning to go, he asked me for severance pay, continued health care benefits, and a professional job recommendation. He was a slow learner.

Another applicant for an upper management position listed on his resume that he spoke fluent Portuguese and Italian. As it turns out, I speak Italian. Part way through the interview I said "Let's continue the interview in Italian." He paled, and mumbled that he had forgotten most of it. He couldn't come up with one phrase. We went back into English. Then one of the staff members visiting from our Brazilian office came by. I asked her to speak to him in Portuguese. Again, he went blank. He had blatantly lied about his language skills.

I could recite many outrageous examples of misrepresentation but will settle for just one more. We hired a middle-aged family man to work in our accounting department. After he was there just a short time, we found that he had written company checks to himself for more than $20,000. When we did a belated background check on him, we found that he had been charged with embezzlement by his previous employer and was awaiting trial. He said he stole the money so that he could spend some quality time with his family before going to jail.

Those examples occurred in the early days of our company. We became smarter with experience. Here are some simple things to do to avoid making such mistakes.

Thoroughly check out job applications including contacting references to be sure that nothing is misrepresented. If educational credentials are claimed, ask the applicant's permission, and then get official transcripts from all institutions he attended. Have the schools send them directly to you.

Make up some very simple tests to roughly measure the applicant's skill level in the specific area in which he claims to have experience. As an example, when hiring engineers we had a 20-question test. One of the test items was for the applicant to draw the wiring diagram for a flashlight. Nothing could be simpler, but not all applicants could do it. Another part of each test was to write on the spot a 50 to 100 word essay on why the applicant thought he was right for the job he had applied for. That sort of test gives a measure of his technical skills as well as his communication skills. You can tell a lot about an applicant by what he writes and how he writes it. As a supplement to your specific home-made tests there are many very good online employee testing services that measure aptitude, personality and skill level. One example of such testing services is found at: **http://www.criteriacorp.com.**

The website offers a battery of employee tests designed by Harvard University psychologists. If you visit the website you can take a sample test yourself to get an idea of how they are done.

Do very thorough background checks on all new hires, particularly on those applying for key positions. If your business has a trained human resources manager he will know how to institute background checks; if not, you should familiarize yourself on the rules and methods for such investigations. A very good Forbes Business article by Mikal E. Belicove titled "The 10 Dos and Don'ts of Conducting Employee Background Checks" can be found at: **www.forbes.com/sites/mikalbelicove/2012/10/26/the-10-dos-and-donts-o-conducting-employee-background-checks/**

It's contains good advice including how to avoid violating job seeker's rights both when conducting such checks and when applying the search results. One key piece of advice: Do not do it yourself; hire a licensed, professional background search firm. They will keep you from doing anything illegal, and you'll get comprehensive results.

Establish a drug-free workplace. Most businesses in the United States conduct mandatory employee testing for drug use. Testing services in your area will be easy to find. When you choose one, make sure their test results are admissible in court should the need ever arise. Have all new hires tested for drug use, and have random testing going forward.

As you add employees, be careful in assigning their titles. In a small start-up company vice presidents and department heads are not needed. Titles casually assigned will quickly become inappropriate and burdensome. From time to time you'll need to bring in new employees at levels above those already in place. Taking a vice president's title away to bring in someone more qualified is painful for everyone. It's much better not to have given the title to him in the first place. Titles are important both to the individuals and to define their roles. But they should be given sparingly. Don't give someone a title hoping he'll grow into the position. Let him do the job well for a good long time, and then offer the title.

In a young, rapid growth company many middle and upper management employees who have possibly given you several years of dedication and personal sacrifice might no longer be qualified for their jobs. It's no fault of theirs or yours. It's actually a great testimony to your collective success. And it doesn't mean start-up employees are less skillful than those who will come on later. In some cases they're actually much more talented. It's just that the required skill sets change as the company grows. Regardless of that, it's a bitter moment when you have to replace, say, your CEO or Production Manager. He was great when you were getting the company started but was unqualified at the $10 million level. The $10 million employee wasn't appropriate when you were just starting out. Now you need him, and you cannot keep them both. Unfortunately the same people whose sweat and blood launched the company very often cannot take it to a higher level. Keep in mind one of them might be you.

Reorganization in a growing company is frequently needed. Don't do it capriciously, but don't hesitate either. Quickly replace or reassign someone who's not up to the job or who's not pulling his load. Hard as it is, you cannot afford to let sentiment block you from keeping the operation lean and efficient. That would sink your ship and everyone in it. I have not always followed this advice myself, and it nearly caused one of my companies to fail. Luckily, much better managers than I came to the rescue in time. You might not be so fortunate.

Always hire the best people you can afford. Never be intimidated by people who are much more talented than you are at many things. Surround yourself with them. Try to be realistic about your own abilities and limitations. Get all the support you can in your own weak areas. Try to accept criticism gracefully.

Please go to the Quiz on page 259.

You should now be able to respond correctly to quiz items 132 through 146.

Product Release

At this point you have legally established your business. Your first product is defined and qualified for use in its intended marketplace. You have a place of business, and have established at least the minimum number of employees and facilities to begin production. You have started making deliverable products. You're almost ready to release them to the marketplace.

Earlier I led you to believe that formal qualification testing was the last step in your test program. That's almost, but not quite, accurate. Putting your product into customers' hands is truly the last "qualification" test, and it's by far the most severe. It's testing that cannot be done in the laboratory, and there's no way to imagine beforehand the many tortures your product will endure in public use. I have been continually amazed and more than a little depressed by customers' involuntary, successful attempts to destroy seemingly robust products.

If your invention is at all fragile or its application easily misunderstood, you should consider starting with a limited release. That will give you a chance to get some practical "use" data before you allow its general distribution. The advantage is that you might learn about some needed modifications before your product is too widely used. If you do a general release and then find problems, you could be faced will a costly recall. You might also alienate a lot of your customers. Because you are an unknown supplier just entering the marketplace, you might lose them forever.

Here's one thing you can do. If the product is a consumer item, you could start by placing a number of units with selected users. Put your friends and family to work once again. Get their reactions to the ease of operation, utility, appearance, and so on. Make up a comment sheet with rankings to check for the product's various functions. Solicit their general input on what they like and don't like about it. Ask them to critically rate it compared to similar products they've used. Find out if they have had any problems. If your product is an industrial item, you could do roughly the same thing by giving out some samples for customer testing. Be sure they are excellent examples of your product, as they will be rigorously tested by your potential customers, and you don't want them to fail.

Upgrading your product based on consumer input never ends. Some problems might take years to surface, but an initial limited release will give you a head start by highlighting the most easily found ones. Once you have taken into account all of the positive and negative input, upgrade your product to incorporate them. Then it's time for a general release to the marketplace.

Earlier in the Product Definition stage you did some research to identify the optimum consumer base for your product. Now you have to find the best ways to get the product in front of the customers. In the course of your research, you looked at all sorts of advertising media to scope out your potential competition and to seek your own product's perfect niche. In short, although it was not your mission at the time, you studied how manufacturers of similar products advertise and sell. Use that information again now.

Pick your top few competitors. Then see how their products are presented. Try to find answers to the following questions:

- Where do they advertise? (Television, radio, magazines, mailers, etc.)
- Do they sell directly to the public?
- Do they have distributors?
- What are their outlets?
- Who are their biggest customers?
- Do they have a sales-representative network?
- What are their biggest-selling items?

- What problems do they seem to have?
- Why are they doing so well/poorly?
- How can I do it better?

If you're at a trade show, go to your competitors' booths. Don't flagrantly misrepresent yourself; just ask whatever questions you want. Where can you find out more about their products? What's the easiest way to buy them? Is there an outlet nearby? Pick up their literature.

When I approach another company's booth, I always introduce myself as one of the competition. Except for those few who are paranoid, they'll usually share as much information with me as they would with any stranger. And you'd be surprised at some of the conversations that follow. For one thing, competitors often have mutually beneficial information they're eager to share, such as pending industry requirements or economic conditions that would adversely affect the marketplace as a whole. Sometimes they gripe about problem customers and kick around ideas of how to tolerate them. Occasionally a top executive or salesmen lets it be known he's ready to change jobs. Often information about other competitors is shared, and so on.

You can make valuable contacts with your competitors this way. And you'll find that a good, ethical competitor isn't a bad thing to have. He'll at times be your ally. On the other hand, voracious, dishonest competitors are an industry pain in the neck. Eventually they are likely to sink themselves, although it might take a disappointingly long time.

Regardless of the information you gather, your most cost-effective sales and advertising methods will probably remain unclear until you've tried various ones. Then you'll be able to sort out the best by developing some statistics on the number and quality of responses you receive from each of them. Like so many other things we've discussed, the approach is very product-specific… and the starting point is determined mainly by informed guesswork.

In addition to whatever other advertising you do, you should have a well-constructed company website... a very modest investment compared to its value.

If you have developed a good product for which there is a demand, and you get it out in front of interested consumers, the orders will begin to flow in. When they do, you must be ready to package it.

Product Packaging and Marking

Packaging is important for nearly all products, and for some it's even more expensive than the product itself. That's particularly true for small off-the-shelf items, or for ones that require special handling. Of course, the packaging must be as carefully designed and economically produced as the product it contains.

The design of packaging is a challenging discipline in itself. Michigan State University's School of Packaging, for example, offers bachelors, masters, and doctorate degrees in the subject.[4] It's worthwhile to take a bit of time to visit their website and see the array of topics they treat. There are also instructions on the site for enrolling in online courses in packaging and package prototyping. The design of packaging is a fertile field for inventors, a fact that can be easily appreciated by paying close attention next time you walk down the aisles of a supermarket.

Attractive, secure packaging will get your product to its destination intact, and will make a good impression. For many products there are legal packaging requirements. Examples are safe transportation of certain goods by air carrier,

4 See the University of Michigan's program at: http://packaging.msu.edu.

and child-proofing dangerous household substances. There are many, many others, resulting in voluminous regulations on the subject.[5] So, you'll have to do some research for your particular product and its transportation modes to be sure you're in compliance. If the transport is interstate or international, destination regional rules will also apply.

Other than those required, there are practical shipping and marking requirements that you should institute yourself. If your product can't be turned upside down, or can't tolerate high temperature, or is otherwise fragile it should be packaged and marked in such a way as to protect it.

Your product and/or its packaging should also be marked with all relevant patent, trademark and copyright notices. Markings should include your company's contact information and web-site address.

A company logo and slogan help distinguish your product from those of your competition, and can add a recognizable, artistic touch. As an example, Nike's simple slash logo and "Just do it." slogan give immediate product recognition in a concise, compact form. When you see that, you know it's Nike. Your name, logo, and slogan may all be subject to trademark protection, and you may be able to get copyright protection for your logo.

Product Pricing

You have made a lot of careful decisions along the way to get to this point. Now you have to make another very essential one: determining the price of your product. Correct pricing is critical. It dictates whether you will eventu-

5 There are myriad Federal Government requirements and guidelines from many agencies such as the Food and Drug Administration, Environmental Protection Agency, Department of Energy, Defense Logistics Agency, Department of Transportation, etc. The voluminous information covers the whole product spectrum. For any certain product, however, it becomes more manageable.

ally make a good profit, just struggle along, or fail completely. Despite its critical importance, however, pricing is often carelessly done. You must take the exercise of pricing your product very seriously. You could start by learning about pricing strategies that are commonly used.

There are many easily found Internet articles that treat product pricing, so it's not hard to develop a minimum understanding of the various ways to do it. Much of the online information is presented by the U.S. Small Business Administration (SBA) in cooperation with its resource partner SCORE. A brief SCORE article titled: "What Drives Your Pricing" is a good starting place.[6] The article outlines five common pricing models, and discusses in some detail the pros and cons of each. If you are new to pricing, the discussion will give you a good introduction. You'll begin to see that there's a lot to consider.

Probably the pricing method most often used for both products and services simply sets price as "unit cost"[7] plus some percentage for profit. While that's conceptually the easiest to envision, it's not always the best and should not be blindly followed. Determining the unit cost is difficult. To be accurate it must include such hard-to-determine factors such as scrap rate, down time, amortization of tooling and equipment, interest on loans, aging of inventory, rent, insurance, warranty returns, quality assurance costs, and on, and on. The method does not take into account any important factors such as competitor's pricing, market demand, or customer satisfaction.

What customers will pay obviously depends on many things having nothing to do with your production cost. Customers don't know what it costs to make your product. They don't care. They know what others charge for similar products, and they know what they're willing to pay. You need to stir

6 www.score.org/blog/2013/ashesh-banerjea/what-drives-your-pricing.

7 Unit cost is just your total actual cost to get the product out the door.

those factors into your pricing equation to get it right. Keep in mind your company's strategic plans are based on profit and revenue projections determined by the price. If you have priced your product on the high side and have some doubts about your ability to sustain it, use a more conservative number for planning purposes. It's better to be surprised on the plus side.

Because pricing is so important, I urge you to first spend some time studying pricing strategies. You can begin on the SCORE website noted above. Once you've done research, do your best to arrive at a price for your product and then seek qualified help to make sure it's optimal. Free professional assistance is waiting for you right there in your own community. Go to your local SCORE office. There you will most likely find an experienced mentor to assist you. Go also to your local SBDC office, and review your strategy with them. Help is there for you. Take advantage of it.

When you have done all of the just-described preparatory steps you're really ready to do business. Your product will start moving out the door. That's a tense and extremely exciting moment, filled with immense satisfaction.

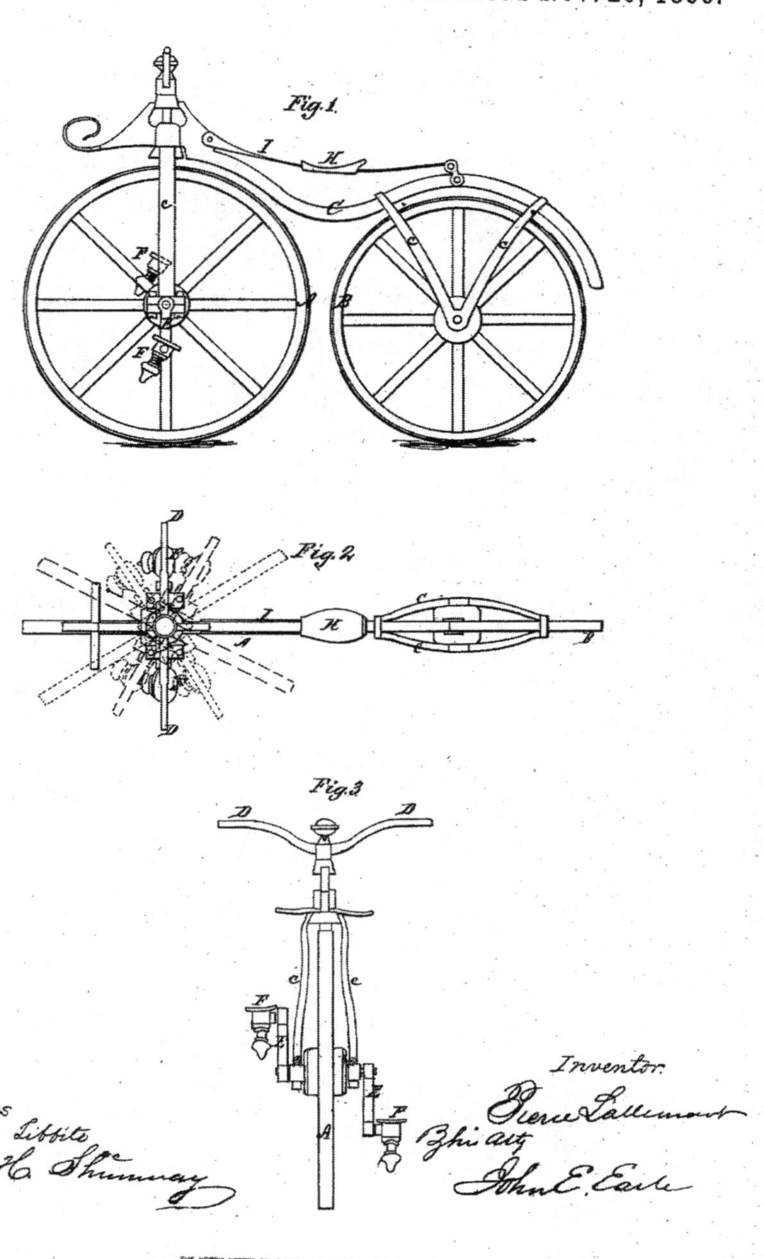
P. LALLEMENT.
VELOCIPEDE.
No. 59,915.
Patented Nov. 20, 1866.
Fig.1.
Fig.2
Fig.3
Witnesses
Inventor:
THE NORRIS PETERS CO., PHOTO-LITHO., WASHINGTON, D.C.

A FEW HARD-EARNED, RANDOM BITS OF ADVICE

As your technology evolves, stay intimately involved so as to not miss a single opportunity to improve it or expand its utility. Do as much as you can yourself. Write procedures for everything critical to your operation. As your methods develop, keep the procedures updated. Archive the outdated procedures. You never know when a step forward might be an error in disguise, and you'll want to go back to the old ways of doing things.

Outfit your business with just the fixed assets you need and no more. Buy for functionality. Be frugal—you'll find many things you must fund, but don't buy anything beyond that. Keep a close eye on your cash flow. Once a company begins to have serious cash problems, it can quickly slide out of control.

Carefully control your inventory. Accumulated inventory might initially show up on your books an asset. But if you cannot turn it over in a timely way, it represents a useless investment of precious funds. And if it becomes obsolete, it's a total loss.

Pay yourself a decent wage. The start-up company might not be able to actually pay you, but you can carry it as a debt owed to you. Or if it's a corporation, take it back as stock or stock options. If the company is eventually successful, you should be paid for your efforts. If not, at least you'll have an accurate measure of your losses.

In all cases be honest and straightforward with your employees, customers, bankers, and vendors. If you have production problems that will cause late delivery, let your customer know as soon as you can. He doesn't need to know how hard you're struggling to catch up or all the details of why. But he does need to know everything that will impact his program.

If you cannot pay a vendor on time, let him know that immediately, and try to work something out. If you're late, don't wait for him to call you; call him first. When you're in a time crunch and really need your vendor's maximum efforts, you don't want him to see you as a potential liability.

Be sure you have adequate banking/credit arrangements in place before you start. When you have a cash emergency, you need a banker who knows he can trust you. Although your banker probably won't have an equity stake in your company, he should be viewed as a partner.

You'll sleep better at night and gain a good reputation if you're up-front and honest 100 percent of the time. Insist on that from your employees as well.

THE END GAME

You have come to the last phase of a long, difficult trail that began with just one fundamentally good idea and matured into a growing company. It's now time to reflect on your goals for the future.

As your start-up business gains traction, formulate your endgame plan. Do you want to be in that business the rest of your career, possibly passing it on to your heirs? Or, do you plan to sell out at some point? These are personal choices made to maximize your satisfaction and rewards, as well as those of your partners and shareholders.

The following discussion is not intended to recommend *how* to sell your business, but instead to give you some thoughts about *when* to sell it. It could help you decide your next move.

Consider the simplistic business cycle model shown in Figure 15. The vertical axis could represent almost any of your company's activities such as sales, technology growth, number of employees, and so on. All of these and other metrics of your activity follow more or less the same sort of life cycle. The horizontal axis is time, and might span 10 years or even more. The life cycle of a start-up technology-based business, can be roughly characterized by three stages: *embryonic*, *midlife*, and *mature*.

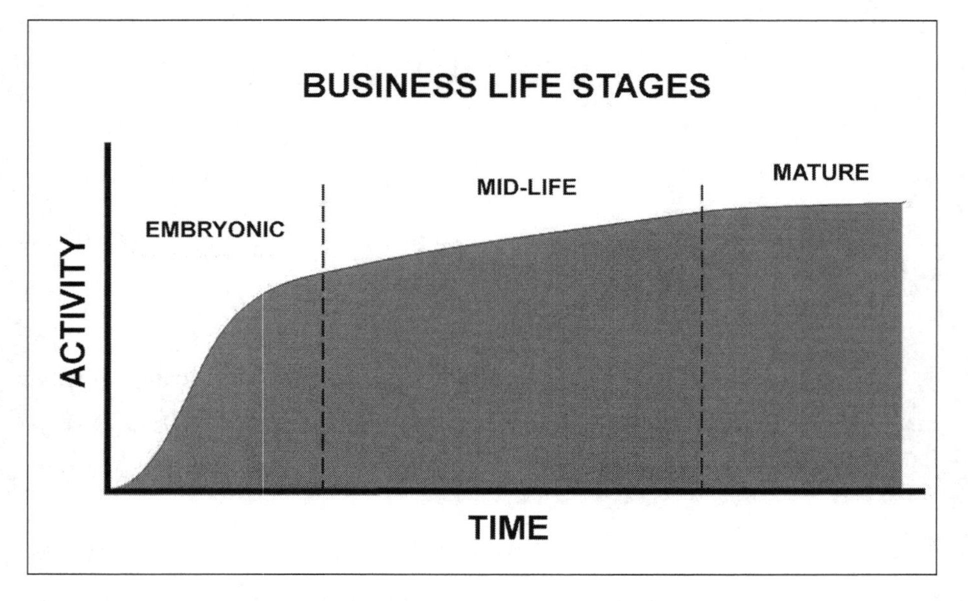

Figure 15

In the *embryonic* stage new-technology applications continue to present themselves, and the first products are taking hold in the marketplace. The business is youthful, growing, and starting to be well defined.

In its *midlife* stage the company has reached its stride, all systems are in place, offshoot products are fewer, and you're down to enhancing profits from sales revenues. You're streamlining your team and systems. You're pedaling as fast as you can.

In the *mature* stage, unless you've continued to produce new innovations, competitors begin to threaten your aging technology, and the company's growth has leveled off or perhaps started to decline.

It's always difficult to generalize, but I believe the optimum time to sell an invention-based start-up company is a few years into it, not more than three or four. By then you should have introduced some product on the market. You will have established the invention's sales potential and efficacy. You'll have a small organization up and running. Your company will be well along into its *embryonic* stage.

At that point an interested, synergistic acquirer would probably evaluate your company more in terms of the intellectual property that you've advanced and protected, than in terms of your earnings. That will be particularly true if your company's sale price is in the range of $1 million to $10 million. Acquirers can spend that much betting on your technology's future without justifying it by your earnings. What you have to sell is the dream of what your promising, substantially proven, technology could become in the hands of its buyers. It is here that the strength of your intellectual property protection will be very important, and it's one more reason to have a professional file your patent application. The breadth and soundness of the protection provided by your patents make up the foundation of the dream you have to sell.

As your business gets well into its midlife stage, an interested buyer will be looking more intently at your financial track record, while still maintaining interest in your protected technology. His evaluation at that point will be largely based on some multiple of your earnings tempered by the earnings' growth rate. Once you get into the midlife stage you'll probably need a few years of stable growth to get the maximum evaluation multiple.[1] It might still vary upward to some extent from the median value of companies like yours due to your protected technology's remaining untapped potential.

It's tempting to go beyond the embryonic stage before selling, because if it's successful, the truth is that the company's earnings will continue to grow. However, the way it's evaluated by a prospective acquirer will also change, and not for the better. So going much beyond the embryonic stage only makes sense if you want to be in for a long haul. I think of the midlife period as a dead-zone for the sale of a company; it hasn't yet made it big, but the dream is starting to fade into reality.

I have been through the experiences of selling both an embryonic company, and a mid-life one. The first sold in about two years, the second in about 20. Both were exciting and very profitable. But I, myself, am more of an inventor than a businessman, and my own skills are more appropriate for a start-up company than for a mature one. Also, I find the challenges of the embryonic-stage more to my liking than those of a more mature company. In the embryonic stage the activities are most often concentrated on solving your company's own technical and formative organizational problems. In the midlife period they are focused more often on solving customer-related problems like delivery schedules, returns for modification, etc. For those reasons, I much preferred the embryonic sale. So, when making your own decision of when to sell, I recommend that you consider how those personal factors apply to you.

1 The earnings' growth rate expressed mathematically is the first derivative of your earnings' history. Potential buyers will likely also look at the second derivative. It takes a few years of data to develop these metrics. They're needed in order to get a good idea of future earnings' potential.

FINAL THOUGHTS

Each of us gets our greatest professional satisfaction in different ways. For me, it's assisting at the birth of new technology, and then firmly launching that technology into the marketplace. And it's being part of a growing young company with all the excitement and camaraderie of an enthusiastic, challenged team. If you've chosen to start a business, you'll see your employees and their families building futures on the jobs that they have in your company. They are a vital part of the community. Your products are out there helping others do things they couldn't do before. Finally seeing your product used, and knowing it makes many people's lives a little better, will give you a wonderful feeling of satisfaction. What could possibly be better.

Finally, I always like to end innovation lectures with Figure 16. It's a patented device for kicking yourself in the butt, and it's just what you deserve if you do not follow your dreams. Go after them.

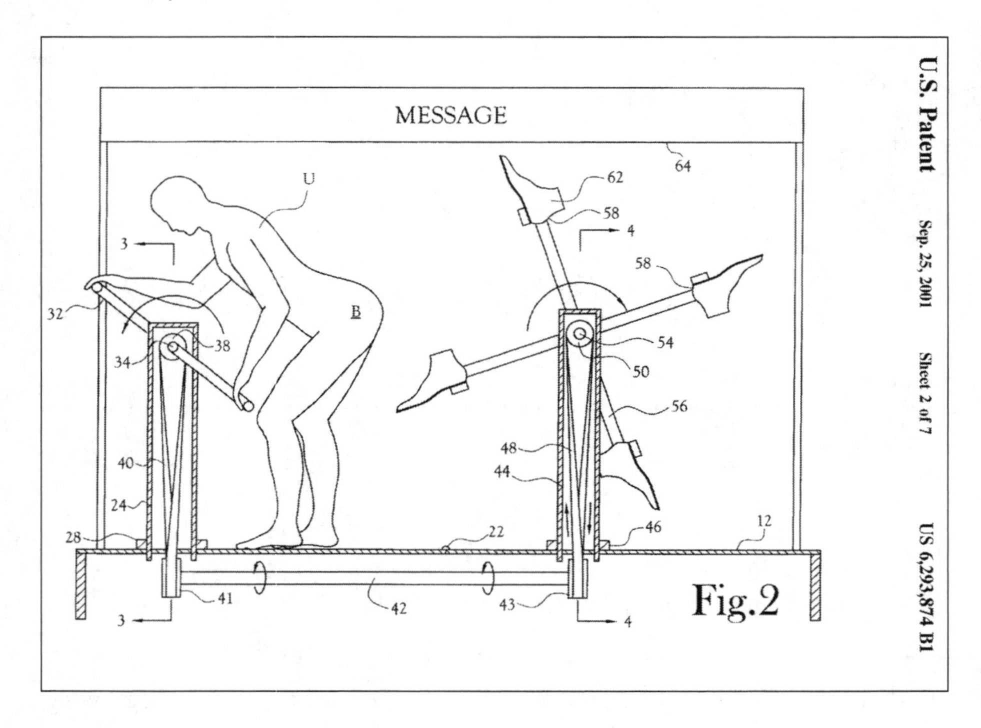

Figure 16

An Endnote to Parents

A successful independent inventor must be both an innovator and an entrepreneur. The book talks a lot about the process of maturing inventions, but much less about being an entrepreneur. It seems to me entrepreneurship is more of an attitude than anything else: it's having the self-confidence and courage to strike out on one's own adventure. That spirit can be encouraged and reinforced through experience. If you want to stimulate your children to become more entrepreneurial, frequently give them some real responsibility outside of the home when they are young. It doesn't have to be much; something as simple as selling Girl Scout cookies or collecting for a charity would help a lot. It gives children the exposure of dealing directly with the public in a somewhat businesslike setting. They have accountability for product and money. They realize repeated rejection is expected; it's not personal, and it doesn't signify failure. And, they feel the joy of accomplishment when they are successful. The lessons learned from those introductory entrepreneurial experiences will stay with the children the rest of their lives, and they will be better for it.

Please go to the Quiz on page 259.

You should now be able to respond correctly to every item on the quiz.

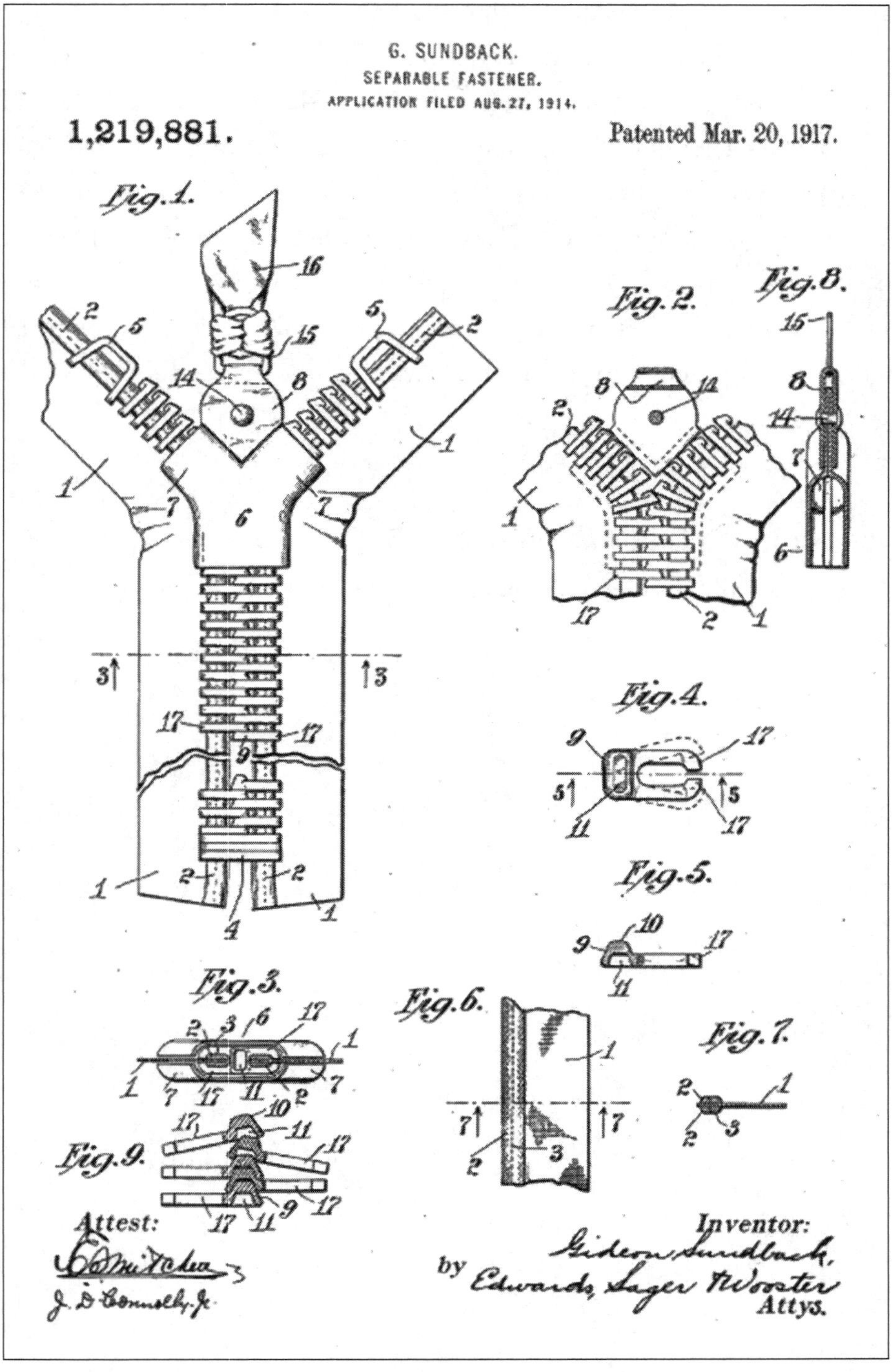

G. SUNDBACK.
SEPARABLE FASTENER.
APPLICATION FILED AUG. 27, 1914.

1,219,881. Patented Mar. 20, 1917.

Fig. 1. Fig. 2. Fig. 3. Fig. 4. Fig. 5. Fig. 6. Fig. 7. Fig. 8. Fig. 9.

Attest:
E. D. Mitchell
J. D. Connolly, Jr.

Inventor:
Gideon Sundback,
by Edwards, Sager & Wooster
Attys.

APPENDIX I: THE AMERICA INVENTS ACT

The American Invents Act (AIA) went into effect on March 16, 2013. It implements fundamental changes to United States patent law. This appendix briefly summarizes some of those changes and their possible implications for you. It also directs you to training videos and other online sources of more comprehensive AIA information. Bear in mind that the AIA rules are still in a state of flux, so rechecking the recommended online sources from time to time is advisable.

Before looking at the four AIA changes most likely to affect you, let us better define the terms "Effective Filing Date," "Prior Art," and "Grace Period."

Effective Filing Date:

A nonprovisional patent application's effective filing date (EFD) is the earlier of its actual filing date, or the filing date of the earliest application from which the invention is entitled to benefit. As two common examples, the EFD could be the filing date of an earlier U.S. provisional application filed by the inventor on the same material, or it could be an earlier Patent Cooperation Treaty (PCT) application the inventor filed in a foreign country naming the U.S. as a country in which patent protection is sought.

Prior Art:

It was previously noted that *prior art* comprises anything that was ever publically known about an invention prior to its EFD. That means if the invention has been patented, described in a printed publication, in public use, on sale, or otherwise available to the public anywhere in the world in any way before that date it is considered prior art, and is no longer patentable.

There are some exceptions to that. Disclosures in any form made by the inventor,[1] or by anyone else who directly or indirectly learned about the disclosed subject matter from the inventor, are not considered prior art if they are made within the one year preceding the application's EFD.[2] If they are made more than one year preceding the EFD they are considered prior art. That one-year interval in which the inventor can openly disclose his invention without losing his ability to legitimately claim it in a U.S. patent is called the "grace period."

1 The term "inventor" as used here includes any and all co-inventors.

2 A good introduction to both prior art and the exceptions just mentioned can be found on the four short USPTO training videos listed here:
Video #1: www.youtube.com/watch?v=wA8_4H156fk
Video #2: www.youtube.com/watch?v=i0qQHTLdRRA
Video #3: www.youtube.com/watch?v=3rL01ItMHvI
Video #4: www.youtube.com/watch?v=YzVrH71c0nc

Grace Period:

During the grace period the inventor is free to publish, advertise, discuss and otherwise reveal his invention in any way he chooses. That freedom comes with some risk. Here's why.

Suppose the inventor publically discloses his invention. Someone learns about it either from him, or from a third party who has heard about it from the inventor or seen a publication issued by the inventor. That person might falsely assert, or even actually think, he has conceived of the invention first. That's not as farfetched as it might seem. It's surprising how often when learning of a new idea we hear someone say: "I thought of that myself years ago." In any case, once the inventor has disclosed his idea he's at risk of someone else claiming it as their own, and filing a patent application before he does. He would probably not even be aware of the bogus application until its publication 18 months after it was filed.

When that happens, he's faced with the unpleasant task of disputing the validity of the filed application. That dispute could take the form of a *Derivation Proceeding,* which will be defined later in this appendix. The inventor possibly would not have the financial means to carry through such a dispute even if he wanted to. If he did launch a dispute, eventually he might be successful in overturning the application, or any patent issuing from it. But he would have to demonstrate that the bogus filer somehow derived the invention either directly or indirectly from him, and that could be difficult to do. What a mess! He would have been far better off not to have disclosed before filing.

#1 AIA Change: First to File

Prior to March 16, 2013, only the *first inventor* of the subject matter disclosed in a U.S. patent application was eligible to be granted a patent on the disclosed invention. After that date, the *first inventor to file* on the subject matter disclosed in a U.S. patent application was eligible to be granted a patent on the disclosed invention. That is an important two-word distinction.

Suppose you have made an invention but are slow to file a patent application on it. Sometime after you made your invention someone else comes up with the same idea, and files a patent application before you do. It is now the second inventor who is eligible to receive a patent, not you. If that situation had occurred before March 2013, you could have challenged the second inventor's right to the patent by demonstrating substantial evidence, such as witnessed test logs or other formal concrete documentation, to substantiate the earlier date of your invention. You can no longer do that. Now, it makes no difference whether you invented it first or not; what matters is which inventor filed first.

As we'll see shortly in a later part of the AIA discussion, post-March, 2013 patents and patent applications are very susceptible to scrutiny and possible overturning of their claims. They must be written with extreme care to discourage attack, or to survive in the event that they are attacked. That need for precision conflicts with the rush to be first-to-file. In ideal circumstances, an inventor should mature his invention as far as practical before filing a patent application. That way, he will be able to make the most comprehensive disclosure on all aspects of his technology and get the broadest claims. But nowadays if he takes the time to do that, he's at risk of having someone else file first, thereby losing everything. It's not a good situation; the two forces are pulling in opposite directions.

#2 AIA Change: Post-grant review

A post-grant review is an action filed by a third party with the object of overturning some or all of the claims of a granted U.S. patent. A post-grant review petition may be filed within nine months after a patent is issued. The petition must present credible evidence that at least one of the patent's claims is unpatentable on any of a number of grounds: for instance, that that the claimed material is not novel; or is obvious; or that the claims are indefinite; or not supported by the specification; that the best mode of use is not disclosed, and so on. In short, most any substantial defect in the patent could trigger a post-grant review, and that review could result in overturning some or all of its claims.

It was recommended in this book's section on patent filing that an individual inventor should not file his own patent unless he has no alternative. To minimize the possibility of attack, it's especially important in this AIA era to have the most carefully written patent possible; that generally means having it done by a professional.

#3 AIA Change: Inter-partes review

An inter-partes review is another action filed by a third party with the intent to overturn one or more of a granted patent's claims. It may be petitioned no sooner than 9 months after the issuance of a patent, and must present reasonable evidence that one or more of the patent's claims is based on prior art that has been disclosed in other patents or printed publications. If the petitioner prevails, the challenged claims in the patent under review will be modified or denied completely.

Much earlier in the book it was mentioned that great care should be taken by the inventor and his representative when writing his patent application to discover and disclose all relevant prior art; hence, the need for adequate patent searching. Failure to do so could easily result in an expensive, time consuming dispute in which the careless inventor might not prevail.

#4 AIA Change: Derivation proceeding

The two previously mentioned review processes are petitioned by third parties. In contrast, derivation reviews are petitioned by an inventor who's filing a patent application (could be you.). If an inventor has substantial credible evidence that someone else has derived the invention from him, and without his authorization improperly filed a patent application on it before he did, he can petition for a derivation review of the suspect application. His object in filing the petition would be to cancel or modify the claims in the earlier application. He must file his petition within one year of the date he first publishes a claim that he believes was improperly claimed in the earlier application. As an example of when a derivation proceeding might be petitioned, please see the earlier Grace Period discussion wherein someone learned of another's invention and improperly filed on it before the true inventor.

THE QUIZ

Quiz answers found on page 271

1. You are encouraged to try to carefully study the simplest things in your world. Why?

2. What are two advantages of starting out with an invention that you can prototype yourself?

3. How should you begin to define your invention?

4. What is the error most commonly made by inexperienced inventors?

5. Why shouldn't you look right away at the inner workings of inventions similar to yours?

6. Why shouldn't you make formal drawings early in the design process?

7. Why should you keep bound, written records of your work?

8. How can the ability to visualize things in the mind's eye be improved?

9. Is all creative thought visual?

10. When should you look in detail at what other designers of similar inventions have done?

11. If you find aspects of inventions similar to yours, and they're in the public domain is it OK to use them?

12. What does the term "public domain" mean?

13. Are aspects of a published patent or patent application not covered in the patent's claims in the public domain?

14. Why should you write a complete assembly procedure for your invention early in the process of its practical evaluation?

15. Why should you write a complete parts list for your invention early in the process of its practical evaluation?

16. Why should you make up a very detailed presentation describing your invention and present it verbally?

17. Name one caution related to making a verbal presentation of your invention.

18. 11 criteria were listed for gauging the commercial potential of your invention. Name six of them.

19. What does the acronym SBA stand for?

20. What does the acronym SBDC stand for?

21. Name two things your local SBDC can do for you.

22. Is SBDC assistance free?

23. What does the acronym BI stand for?

24. Are BI services free?

25. Do BI's and SBDC's work together?

26. What sort of organization is SCORE?

27. What is one major advantage of SBIR funding?

28. How long should you wait before seeking development funds that dilute your business ownership?

29. What does the term "intellectual property" encompass?

30. How is intellectual property protected?

31. Why should you check your employment agreement and intellectual property employment rules in your state before investing much time and money in your invention?

32. What are "shop rights?"

33. What condition must be met for one to be considered a legitimate co-inventor?

34. Can a financial backer who has not contributed intellectually to your invention be considered a co-inventor?

35. If you have a co-inventor who has made an intellectual contribution resulting in only one of your patent's claims, does he have as much right to the entire patent as you have?

36. To whom can you safely reveal your pre-patent confidential information?

37. What does the acronym USPTO stand for?

38. Does a U.S. patent give you the exclusive right to sell your invention?

39. An analogy is often made between a patent and what other legal document?

40. Approximately what fraction of patent applications submitted to the USPTO result in the granting of patents?

41. Name the three types of patents.

42. What is the most desirable type of patent?

43. Name three conditions that must be met in order to qualify for a utility patent:

44. What does the acronym EFD stand for?

45. What is "prior art?"

46. Does the USPTO check to make sure the invention represents a good idea before granting a patent?

47. What is the lifetime of a utility patent?

48. When are utility patent maintenance fees due?

49. What happens if maintenance fees are not paid on time?

50. Who other than the inventor can apply for a patent?

51. What does a design patent cover?

52. Is the lifetime of a design patent 14 years after the application effective filing date?

53. Is it possible to be granted a design patent and a utility patent on the same invention?

54. When are maintenance fees for design patents due?

55. What type of plants do plant patents cover?

56. What is a provisional patent?

57. How soon after filing a provisional patent application must the corresponding nonprovisional application be filed to benefit from the provisional's earlier filing date?

58. When is a provisional patent application published?

59. Name two reasons to file a provisional patent application on your invention.

60. If a nonprovisional patent application is not filed within one year of filing a corresponding provisional application, can a new provisional application be filed on the same subject matter?

61. Does a provisional patent application allow the inventor to mark the subject invention with the notice: Patent Pending?

62. Give one example of a well-known trademark.

63. Can you establish your rights to a trademark simply by using it?

64. Name one reason to register your trademark with the USPTO?

65. What is the strongest type of trademark?

66. What do copyrights protect?

67. Are copyrights registered with the USPTO?

68. Are copyrights obtained automatically when the work is published?

69. What is a "trade secret?"

70. Name one situation in which trade secret protection is superior to patent protection.

71. What happens if your trade secret is accidently published?

72. What is the difference between a dependent claim and an independent patent claim?

73. What does the "claims" section of a granted patent define?

74. Why is it important for the inventor to assist in defining a patent application's claims?

75. Do wordy patent claims offer narrower or broader protection than succinct claims?

76. When is a provisional patent application reviewed by the USPTO?

77. About how much does it cost to have an attorney file an uncomplicated nonprovisional patent application?

78. What is a patent application's "first office action?"

79. About how long does it take from the time a patent application is filed until it is granted?

80. Once your patent application is published you are entitled to reasonable royalties from an infringer providing two conditions are satisfied. What are they?

81. Name some good practical reasons for performing a DIY patent search.

82. What are the main things to look for when doing a DIY patent search?

83. When doing a DIY patent search is it necessary to search published applications as well?

84. Which claim in a patent most often gives the broadest coverage?

85. What is the best way for an individual inventor to perform patent searches?

86. What information does the "Legal Events" section of patents viewed on the Google website contain?

87. What is the "Official Gazette?"

88. What is a "claims infringement analysis?"

89. What is a "claims analysis table?"

90. How many of the features of another patent's claim must the comparable claim of your patent have in order to infringe it?

91. Can unclaimed features disclosed in other patents affect the breadth of your allowed claims?

92. Are professional patent search firms registered by the USPTO?

93. About how much does a professionally performed patent search accompanied by an attorney's opinion cost?

94. In which state is the USPTO located?

95. Where is the National Inventors Hall of Fame located?

96. May any inventor file his own patent?

97. In what circumstances is it advisable to file your own patent?

98. Which two sorts of professionals are registered to file patent applications with the USPTO?

99. Are attorneys not registered with the USPTO able to represent you in disputes before the USPTO?

100. Why should an individual inventor bother learning about the format and content of patents?

101. What are the advantages of your writing the first draft of your patent application?

102. What does the acronym "PCT" stand for?

103. What sort of protection does a PCT patent application provide?

104. What is the effective lifetime of a PCT patent application?

105. What are some advantages of filing a PCT patent application?

106. Once you filed a patent application, what may you do with your invention?

107. Are fraudulent firms disguising themselves as patent marketing and development firms common?

108. Name a couple of facts about their business practices an invention promotional firm is legally required to disclose before it enters into a contract with you.

109. What's one way to learn if any complaints have been registered against an invention promotional firm?

110. What are the four most common ways to cash in on an invention?

111. What does full assignment of a patent mean?

112. Name one circumstance in which you might be obliged to assign your patent rights to another person or entity.

113. Is it possible to make a partial assignment of your patent rights?

114. List three of the many considerations a partial assignment contract should contain.

115. In a license agreement, which party is the licensor, and which is the licensee?

116. What does patent licensing mean?

117. Are you required by law give a licensee of your intellectual property a warranty that your patented technology doesn't infringe someone else's rights?

118. Can you license your intellectual property to more than one entity at a time?

119. Why might a candidate licensee not want to receive any proprietary information from you?

120. What is the "not invented here" syndrome?

121. Who might the best person in a candidate licensee's business be to first present your licensing proposal?

122. When you present your licensing proposal to a potential licensee is it wise to show your prototypes?

123. Why would a trade show be a good place to look for a potential licensee?

124. About how long should a license option agreement last?

125. What is the purpose of a license option agreement?

126. What is the difference between sub-licensing an intellectual property license agreement and transferring the agreement?

127. What is a license royalty?

128. What are royalty payments usually based upon?

129. What are typical royalty percentages?

130. What is a "minimum annual royalty?"

131. Should the licensor or the licensee retain improvement rights to the licensed intellectual property?

132. What is one disadvantage of entering into a partnering agreement?

133. What does "Proof of concept" testing do?

134. What are the advantages of testing a device's individual concepts using simple mock-ups as opposed to using full prototypes?

135. Why should you participate yourself in the proof-of-concept testing of your invention?

136. When a proof-of-concept test fails, what's the first thing to suspect?

137. What role does market research play in product definition?

138. What are two ways that functional prototypes can serve you?

139. What is the fundamental difference between proof-of-concept testing and qualification testing?

140. How long after your start-up business is established should you write your business plan?

141. What is a SWOT analysis?

142. What does the acronym SWOT stand for?

143. Why are explosive sales a business threat?

144. How can the local Chamber of Commerce help you when choosing a new business site?

145. What is the website: www.BusinessUSA.gov intended to do?

146. How can you be reimbursed up to 50% for costs to provide on-the-job training for employees?

147. Why is it a good idea to start with a limited product release?

148. Name a couple of reasons why your product might need special packaging:

149. Why is correct pricing of your product so important?

150. What is the "unit cost" method of pricing?

151. What free, local resource is useful to help you determine the price of your product?

152. Why is it a good idea to archive outdated manufacturing procedures?

153. What are the three lifetime stages of a company?

154. In which of the lifetime stages of a company would its sale represent selling a "dream?"

155. What would a potential business acquirer be looking at most carefully in estimating a mid-life company's value?

156. Why is it important to give your children some real businesslike responsibility outside of the home?

QUIZ ANSWERS

1. To improve your understanding of how things function and how they are made. (Page 31)

2. Maximizes your hands-on involvement and minimizes expense. (Page 33)

3. Set down your design goals. (Page 34)

4. Disregard for cost. (Page 35)

5. It stifles your own creativity. (Page 35)

6. It constrains your creativity. (Page 36)

7. It helps you remember what you've done and could help resolve ownership disputes. (Page 36)

8. With practice. (Page 38)

9. No, some is verbal. (Page 38)

10. When you can do no more to advance your own invention through your own thought alone. (Page 46)

11. Yes. (Page 46)

12. Things that are already publically known and are not privately held are said to be in the public domain. (Page 46)

13. Yes. (Page 46)

14. The exercise can reveal assembly difficulties and design problems more easily solved at an early stage of development. (Page 46)

15. The exercise can give you insight to what the eventual sales price will be. (Page 47)

16. You will learn something more about your invention this way. (Page 47)

17. You might acquire an unwanted co-inventor, or, your audience might not respect the confidentiality of your disclosure. (Page 47)

18. Any six of the following: demand, stand-out from competition, protection against knock-offs, necessary development resources, manufacturability, user friendly, affordability, enduring market, barriers to market entry, legal test requirements, and simple beauty. (Page 53 through 59)

19. U.S. Small Business Administration. (Page 62)

20. Small Business Development Center. (Page 62)

21. Help you create your business plans, help organize patent protection for your invention. (Page 62)

22. Yes. (Page 62)

23. Business Incubator. (Page 63)

24. Usually no. (Page 63)

25. Frequently. (Page 64)

26. It's a public service organization dedicated to offering free, professional, business advice. (Page 65)

27. It requires no dilution of your business ownership. (Page 65)

28. As long as you possibly can. (Page 69)

29. It includes innovations such as inventions, writings, music, works of art, and other products created in one's mind. (Page 71)

30. By patents, copyrights, trade secrets, and contracts. (Page 71)

31. Your ownership might be compromised by conditions of your employment. (Page 72)

32. Shop rights are rights to ownership in your invention an employer might be entitled to if you have used some of his resources in creating or advancing your invention. (Page 73)

33. He must have made a substantial intellectual contribution to the invention that results in at least one of the issued patent's claims to uniqueness. (Page 73)

34. No, adding financial support does not qualify him as a co-inventor. (Page 74)

35. Yes, in the absence of any other contracts, all named inventors on an invention have equal rights. (Page 75)

36. Your USPTO-registered patent agent or attorney. (Page 77)

37. The United States Patent and Trademark Office. (Page 77)

38. No, It defines your invention and gives you a tool for defending it, but it does not guarantee that you can sell it legally. (Page 79)

39. A land deed. (Page 80)

40. Two-thirds. (Page 80)

41. Utility, design, and plant patents. (Page 81)

42. Utility patent. (Page 81)

43. Useful, novel, non-obvious. (Page 82)

44. Effective filing date. (Page 82)

45. Anything that has ever been publically known about an invention other than that disclosed by the inventor within the year preceding its EFD, is known as prior art. (Page 82 and Appendix I)

46. No. (Page 82)

47. Twenty years from the effective filing date. (Page 84)

48. In the middle of the third, seventh, and eleventh years after the patent grant date. (Page 84)

49. The patent expires. (Page 84)

50. Anyone, such as a qualified employer, who has a proprietary interest in the invention. (Page 84)

51. An object's new and non-obvious ornamental, non-functional features. (Page 85)

52. No, it is 14 years after the patent grant date. (Page 85)

53. Yes, one to cover how it works and the other to cover how it looks. (Page 85)

54. Never. Maintenance fees are not required for design patents. (Page 85)

55. New, distinct asexually produces varieties. (Page 85)

56. There is no such thing. A provisional patent applicant does not result in a provisional patent. (Page 87)

57. Within one year. (Page 87)

58. A provisional patent application is never published. (Page 87)

59. The provisional application's principal advantages are that it lets the inventor get something formal on record and buys him a little time to advance his program before laying out the cash for a nonprovisional filing. (Page 87)

60. Yes. (Page 87)

61. Yes. (Page 88)

62. McDonald's Golden arches, as one of very many. (Page 89)

63. Yes, if it's used correctly. (Page 89)

64. The ability to bring action regarding the trademark in federal court, amongst others. (Page 89)

65. Highly stylized, it makes it more likely that it will not infringe someone else's trademark. (Page 90)

66. Copyrights protect creative works such as writings, music, and works of art. (Page 91)

67. No, they are registered with the U.S. Copyright Office. (Page 91)

68. Yes. (Page 91)

69. Any proprietary material describing the invention, including manufacturing procedures, drawings, etc. that is held in confidence by the originator. (Page 91)

70. When the matter to be protected is a method, process, or recipe which is difficult to reproduce by others. (Page 92)

71. It is no longer a secret, and no longer guarantees you any protection against others copying it. (Page 93)

72. Dependent claims refer back to and depend upon independent ones. (Page 108)

73. It defines the scope of the patent's protection. (Page 108)

74. The inventor is the person who should be most aware of the new technology's advantages and potential applications. (Page 109)

75. As a rule-of-thumb, the more words a patent's claim contains, the less it protects. (Page 112)

76. It will not be considered until the nonprovisional application is examined, and then it will be viewed to see if it supports the nonprovisional's content. (Pages 112)

77. Between $10,000 and $15,000. (Page 113)

78. A first office action is the USPTO's initial response to a filed nonprovisional patent application. (Page 113)

79. Currently 22 months. (Page 114)

80. You must formally put the infringer on notice, and at least one claim from your granted patent must be infringed. (Page 114)

81. Get good ideas for your own invention's applications; learn about competitive products; find conflicts with prior art; helps distinguish your invention from others; learn if others failed and why. (Page 121)

82. Existing art that could render your invention not-unique or that would make it obvious. (Page 123)

83. Yes. (Page 123)

84. The first claim. (Page 123)

85. Computer key-word searches on Google. (Page 124)

86. Any legal actions such as assignments, expiration, and maintenance fee payment history. (Page 126)

87. It is a weekly USPTO online publication of granted and expired patents. (Page 128)

88. It is an analysis to determine if your invention trespasses on the rights of someone else's patent claims. (Page 128)

89. It is a systematic way to compare the claims language of two patents to find differences and similarities. (Page 131)

90. All of them. (Page 133)

91. Yes, all unclaimed elements of published patents are in the public domain, and are prior art. (Page 137)

92. No. (Page 140)

93. Around $1500. (Page 141)

94. Virginia. (Page 141)

95. In the same building as the USPTO facility, in Virginia. (Page 142)

96. Yes. (Page 143)

97. Only if there simply is no alternative. (Page 144)

98. USPTO-registered patent agents and patent attorneys. (Page 144)

99. No. (Page 145)

100. Both to understand how to search patents and to assist in filing them. (Page 147)

101. It makes it easier for the attorney or agent to understand the invention and what you see as its most important features. (Page 147)

102. Patent Cooperation Treaty. (Page 147)

103. It gives the inventor interim patent protection in the 148 industrialized countries which now participate in the treaty. (Page 147)

104. 30 months from the effective filing date of the corresponding U.S. patent application. (Page 148)

105. One advantage is that it gives you interim protection in most all industrialized countries for a modest filing expense. Another advantage is that it allows deferral of some filing fees by granting more time to decide

whether or not to file nonprovisional U.S. or foreign patent applications and to decide in which foreign countries to seek more lengthy protection. (Page 149)

106. You could lease, sell, or further develop it. (Page 150)

107. Yes. (Page 151)

108. Any two of the following constitutes a correct answer: How many inventions it has evaluated; how many received positive evaluations; total number of customers; how many customers made a net profit; how many customers licensed inventions through them; names and addresses of all invention promotion companies it's been affiliated with over the past ten years. (Page 154)

109. Visit the complaints section of the USPTO website. (Page 156)

110. Licensing; assigning; partnering; starting a business to sell the invented product. (Page 158)

111. It means transferring all of your rights to your patent to someone else. (Page 159)

112. Under terms of your employment agreement. (Page 160)

113. Yes. (Page 161)

114. Check the long list on page 162 for the correct answers. (Page 162)

115. The licensor is the one granting the license; the licensee is the one getting the license. (Page 165)

116. It means granting another person or entity the right to use you intellectual property. (Page 165)

117. No, and you shouldn't do so. (Pages 165)

118. Yes, license contracts can be whatever the parties agree upon. (Page 166)

119. He might run the risk of litigation from you downstream. (Page 168)

120. It's the tendency of engineers in a licensee's company to reject outside ideas. (Page 168)

121. The marketing manager. (Page 168)

122. Yes. (Page 170)

123. It's a great place to meet industry leaders. (Page 171)

124. Sixty days or so. (Page 172)

125. It gives the potential licensee time to make a decision about licensing your intellectual property. (Page 172)

126. A licensee who sub-licenses your agreement remains responsible to you; one who transfers it does not. (Page 176)

127. It is a payment to the licensor for the right to use his intellectual property. (Page 177)

128. Some percentage of sales of the licensed products. (Page 177)

129. They typically fall within the range of 3% to 6% of sales. (Page 178)

130. It is a guaranteed minimum amount that a licensee pays each year to retain the license. (Page 180)

131. Licensor. (Page 182)

132. Loss of flexibility. (Page 184)

133. It proves the validity of the fundamental concepts on which an invention is based. (Page 202)

134. One is that testing fewer things at once makes it easier to pinpoint a failure's source. Another is that it's generally quicker, easier, and much less expensive. (Page 203)

135. You get valuable hands-on experience with your invention's development. (Page 208)

136. The test apparatus. (Page 209)

137. It allows the product to be tailored for the most immediate and profitable markets. (Page 211)

138. One is for qualification testing; the other is to show off your product. (Page 215)

139. The former demonstrates the soundness of the invention's various concepts. The latter shows that the product, in the form in which it will be sold, meets all of its performance requirements. (Page 217)

140. It should be written prior to starting up. (Page 223)

141. It is a systematic way to analyze your business. (Page 224)

142. Strengths, Weaknesses, Opportunities, and Threats. (Page 224)

143. They can cause severe cash flow and staffing problems. (Page 225)

144. By providing comprehensive regional information and permitting. (Page 227)

145. To disseminate core information regarding the Federal Government's programs and services for small businesses. (Page 228)

146. By hiring them through an American Job Center. (Page 230)

147. To get some practical use data before its general distribution. (Page 235)

148. If it's fragile; cannot stand heat; must be kept right-side up. (Page 239)

149. It dictates whether you will make a profit or fail. (Page 240)

150. It takes the sum of all costs incurred in producing one unit of a product, and then adds a percentage of that for profit. (Page 240)

151. SCORE (Page 241)

152. You might want to go back to an earlier way to do things. (Page 243)

153. Embryonic, mid-life, and mature. (Page 246)

154. Embryonic (Page 247)

155. Financial track record. (Page 247)

156. It helps them develop an entrepreneurial attitude. (Page 250)

ABOUT THE AUTHOR

Award winning Physicist/Oceanographer Dr. James L. Cairns is a lifelong inventor with more than 60 issued and pending U.S. Patents. Most of his technical contributions have been components for deep-ocean power and communications networks. They enabled offshore oil and gas operations to move into the very deep sea, greatly expanding the global area accessible for offshore energy production. His systematic approach to inventing, based on over five decades of successful experience, provides a useful and inspiring model for other independent inventors. He lives in Ormond Beach, Florida, but spends nearly half of his time living and working in a renovated thirteenth-century monastery near the city of Urbino, Italy.

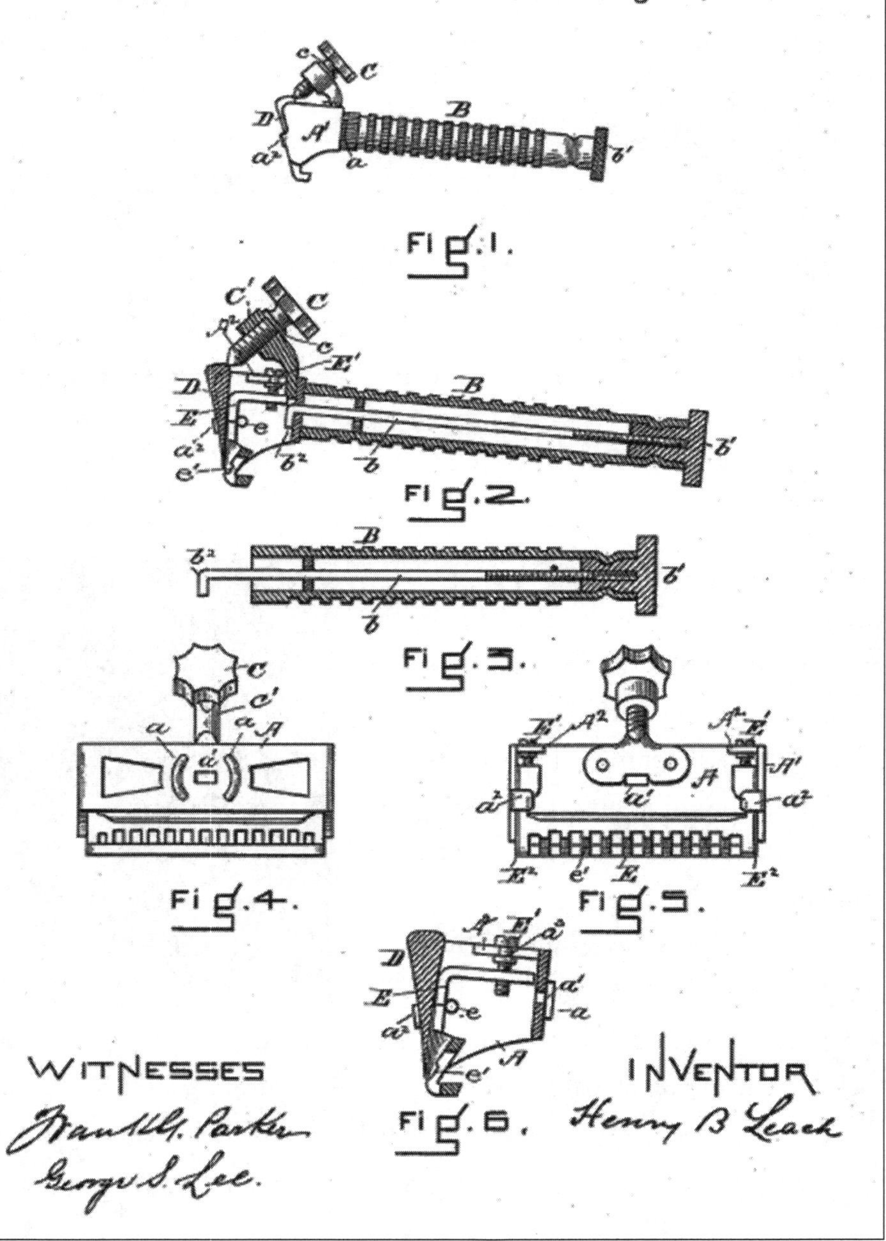
(No Model.)
H. B. LEACH.
SAFETY RAZOR.
No. 434,187.
Patented Aug. 12, 1890.
FIG.1.
FIG.2.
FIG.3.
FIG.4.
FIG.5.
FIG.6.
WITNESSES
INVENTOR
Henry B Leach

INDEX

E

F

G

H

I

Q

R

S

T

U

V

W